T0238981

Convergence Methods for Double Sequences and Applications

M. Mursaleen · S.A. Mohiuddine

Convergence Methods for Double Sequences and Applications

M. Mursaleen
Department of Mathematics
Aligarh Muslim University
Aligarh, Uttar Pradesh, India

S.A. Mohiuddine
Department of Mathematics
King Abdulaziz University
Jeddah, Saudi Arabia

ISBN 978-81-322-2925-4   ISBN 978-81-322-1611-7 (eBook)
DOI 10.1007/978-81-322-1611-7
Springer New Delhi Heidelberg New York Dordrecht London

© Springer India 2014
Softcover reprint of the hardcover 1st edition 2014
This work is subject to copyright. All rights are reserved by the Publisher, whether the whole or part of
the material is concerned, specifically the rights of translation, reprinting, reuse of illustrations, recitation,
broadcasting, reproduction on microfilms or in any other physical way, and transmission or information
storage and retrieval, electronic adaptation, computer software, or by similar or dissimilar methodology
now known or hereafter developed. Exempted from this legal reservation are brief excerpts in connection
with reviews or scholarly analysis or material supplied specifically for the purpose of being entered
and executed on a computer system, for exclusive use by the purchaser of the work. Duplication of
this publication or parts thereof is permitted only under the provisions of the Copyright Law of the
Publisher's location, in its current version, and permission for use must always be obtained from Springer.
Permissions for use may be obtained through RightsLink at the Copyright Clearance Center. Violations
are liable to prosecution under the respective Copyright Law.
The use of general descriptive names, registered names, trademarks, service marks, etc. in this publication
does not imply, even in the absence of a specific statement, that such names are exempt from the relevant
protective laws and regulations and therefore free for general use.
While the advice and information in this book are believed to be true and accurate at the date of publication, neither the authors nor the editors nor the publisher can accept any legal responsibility for any
errors or omissions that may be made. The publisher makes no warranty, express or implied, with respect
to the material contained herein.

Printed on acid-free paper

Springer is part of Springer Science+Business Media (www.springer.com)

Preface

This short monograph is the first one to deal exclusively with the study of almost convergence and statistical convergence of double sequences. In almost every branch of science, engineering, medical science, life science, social science, and business, we often come across some sort of sequences. So the study of sequences or sequence spaces in broader sense becomes a very important topic to deal with such situations. To get or check the limit of a sequence, there are several ways and methods. Sometimes, the limit of a sequence may not exist, but its generalized limit or weak limit may exist to serve our purpose. The notion of almost convergence is one of the most useful notions to obtain a weak limit of a bounded nonconvergent sequence. There is another notion of convergence, known as the statistical convergence, which was introduced by H. Fast, who attributed this concept to Hugo Steinhaus. Fast introduced an extension of the usual concept of sequential limit, which he called statistical convergence. In 1953, this concept arose as an example of "convergence in density," which was also studied as a summability method. Even unbounded sequences can be dealt with this method.

In Chap. 1, we give an overview of almost convergence, statistical convergence, and some related methods for ordinary (single) sequences. In Chap. 2, we discuss the notion of almost convergence and almost Cauchy for double sequences. Some more related spaces for double sequences, associated sublinear functionals, and various inclusion relations are also studied. The theory of sequence spaces gave rise to another important topic, matrix transformations, where we study the methods to transform one sequence space into the same or another sequence space. Such a study of matrix transformations generalizes some special methods of summability. These methods are studied in Chap. 3, in which we characterize the classes of four-dimensional almost conservative, almost regular, strongly regular, and almost strongly regular matrices. In Chap. 4, we study the concept of absolute almost convergence for double sequences and use this notion to characterize the absolute almost conservative and absolute almost regular matrices. In Chap. 5, we define different cores of double sequences and establish various core theorems analogous to the well-known Knopp's core theorem. In Chap. 6, we give some applications of almost convergence of double sequences to prove Korovkin-type approximation theorems for functions of two variables through different sets of test functions and also show that these results are more applicable. Chapter 7 is devoted to the study of statistical convergence of double sequences, and in Chap. 8, we apply this to statistical approximation of positive linear operators. The last chapter is devoted to the study of convergence of double series and describe various convergence tests. We give many interesting examples. For the convenience of the reader, all chapters of this book are written in a self-contained style.

This brief monograph relies mostly on both authors' recent research work as well as other eminent authors' work. Advanced courses can be taught out of this short book. All necessary background and motivation are given per chapter. Besides applications in approximation theory, the presented results are expected to find applications in many other areas of pure and applied mathematics, such as mathematical analysis, probability, fixed point theory, statistics, etc. As such, this brief monograph is suitable for researchers, graduate students, and seminars on the above subjects.

This book was basically started when the first author visited King Abdulaziz University, Jeddah, Tabuk University, Tabuk, Yildiz Technical University, Istanbul, and University Putra Malaysia during 2011 and 2012, to whom he is greatly thankful for providing him kind hospitalities during the stay at these institutions. Both the authors would like to thank their respective families for their moral support during the writing of this monograph.

Aligarh, India   M. Mursaleen
Jeddah, Saudi Arabia   S.A. Mohiuddine

Contents

Chapter 1
Almost and Statistical Convergence of Ordinary Sequences: A Preview

In this chapter, we recall the notion of almost convergence and statistical convergence for single sequences $x = (x_k)$. We present here a brief survey on developments of almost convergence, statistical convergence, and some related methods, e.g., absolute almost convergence and strong almost convergence for single sequences.

1.1 Introduction

Let c and l_∞ denote the spaces of all convergent and bounded sequences $x = (x_k)$, respectively. Note that $c \subset l_\infty$. In the theory of sequence spaces, an application of the well-known Hahn–Banach extension theorem gave rise to the notion of Banach limit, which further leads to a beautiful concept of almost convergence. That is, the lim functional defined on c can be extended to the whole of l_∞, and this extended functional is known as the Banach limit [8]. In 1948, Lorentz [63] used this notion of weak limit to define a new type of convergence, known as the almost convergence. Since then, a huge amount of literature has appeared concerning various generalizations, extensions, and applications of this method. In [109], Raimi gave a slight generalization of the Banach limit, which is known as the invariant mean or σ-mean and known as the σ-convergence. Another generalization of almost convergence was given by Stieglitz [121] by using a sequence of infinite matrices, known as the F_B-convergence. Absolute and strong analogues of these methods have also been studied by various authors (cf. [28–33, 64, 65, 84–87]). In this chapter, we provide to the readers a good account of the developments of the notion of almost convergence and some related methods. More information can be found in [1, 27, 42, 55, 89, 118].

We shall write ω for the set of all complex sequences $x = (x_k)_{k=0}^{\infty}$. Let ϕ and c_0 denote the sets of all finite and null sequences, respectively. We write $l_p := \{x \in \omega : \sum_{k=0}^{\infty} |x_k|^p < \infty\}$ for $1 \leq p < \infty$. By e and $e^{(n)}$ $(n \in \mathbb{N})$ we denote the sequences such that $e_k = 1$ for $k = 0, 1, \ldots$, and $e_n^{(n)} = 1$ and $e_k^{(n)} = 0$ $(k \neq n)$. For any sequence $x = (x_k)_{k=0}^{\infty}$, let $x^{[n]} = \sum_{k=0}^{n} x_k e^{(k)}$ be its n-section.

M. Mursaleen, S.A. Mohiuddine, *Convergence Methods for Double Sequences and Applications*, DOI 10.1007/978-81-322-1611-7_1, © Springer India 2014

Note that c_0, c, and l_∞ are Banach spaces with the sup-norm $\|x\|_\infty = \sup_k |x_k|$, and l_p $(1 \leq p < \infty)$ are Banach spaces with the norm $\|x\|_p = (\sum |x_k|^p)^{1/p}$, while ϕ is not a Banach space with respect to any norm.

A sequence $(b^{(n)})_{n=0}^\infty$ in a linear metric space X is called a *Schauder basis* if for every $x \in X$, there is a unique sequence $(\beta_n)_{n=0}^\infty$ of scalars such that $x = \sum_{n=0}^\infty \beta_n b^{(n)}$. A sequence space X with a linear topology is called a *K-space* if each of the maps $p_i : X \to \mathbb{C}$ defined by $p_i(x) = x_i$ is continuous for all $i \in \mathbb{N}$. A K-space is called an *FK-space* if X is complete linear metric space; a *BK-space* is a normed *FK*-space. An *FK*-space $X \supset \phi$ is said to have *AK* if every sequence $x = (x_k)_{k=0}^\infty \in X$ has a unique representation $x = \sum_{k=0}^\infty x_k e^{(k)}$, that is, $x = \lim_{n \to \infty} x^{[n]}$, and X is said to have *AD* if ϕ is dense in X. It is evident that *AK* implies *AD*.

1.2 Almost Convergence

1.2.1 Definitions and Examples

Definition 1.1   A linear functional L on l_∞ is said to be a *Banach limit* if it has the following properties:

(i) $L(x) \geq 0$ if $x \geq 0$,
(ii) $L(e) = 1$, and
(iii) $L(Sx) = L(x)$, where S is a shift operator defined by $(Sx)_n = x_{n+1}$.

Definition 1.2   A bounded sequence $x = (x_k)$ is said to be *almost convergent* to the value ℓ if all its Banach limits coincide, i.e., $L(x) = \ell$ for all Banach limits L.

Theorem 1.3   *Let β denote the set of all Banach limits. If $L \in \beta$ and $x \in l_\infty$, then*

$$\liminf x_n \leq -p(-x) \leq L(x) \leq p(x) \leq \limsup x_n, \tag{1.1}$$

where $p(x)$ is a sublinear functional on l_∞ defined by

$$p(x) = \inf_{n_1, n_2, \ldots, n_p} \limsup_{k \to \infty} \frac{1}{p} \sum_{i=1}^p x_{n_i + k}. \tag{1.2}$$

Lorentz [63] established the following characterization.

Theorem 1.4   *A sequence $x = (x_k)$ is almost convergent to the number ℓ if and only if $t_{mn}(x) \to \ell$ as $m \to \infty$ uniformly in n, where*

$$t_{mn} = t_{mn}(x) = \frac{1}{m+1} \sum_{i=0}^m x_{n+i}. \tag{1.3}$$

The number ℓ is called the generalized limit of x, and we write $\ell = f\text{-}\lim x$. We denote the set of all almost convergent sequences by f, i.e.,

$$f := \left\{ x \in l_\infty : \lim_m t_{mn}(x) = \ell, \text{ uniformly in } n \right\}. \qquad (1.4)$$

The sequences that are almost convergent are said to be summable by the method f, i.e., by $x \in f$ we mean that x is almost convergent and $f\text{-}\lim x = L(x)$.

Examples 1.5

(i) For $z \in \mathbb{C}$ on the circumference of $|z| = 1$, $L(z^n) = 0$ everywhere except for $z = 1$. The assertion follows immediately from

$$\left| \frac{1}{k}(z^n + z^{n+1} + \cdots + z^{n+k-1}) \right| = \left| z^n \frac{1 - z^k}{k(1 - z)} \right| \leq \frac{2}{k(1 - |z|)}.$$

It is easy to see that the geometric series $\sum z^n$ for $|z| = 1$, $z \neq 1$, is almost convergent to $1/(1 - z)$. Hence, it follows that the Taylor series of a function $f(z)$ that is regular for $|z| < 1$ and has simple poles on $|z| = 1$ is almost convergent at every point of the circumference $|z| = 1$ with the limit $f(z)$.

(ii) A periodic sequence (x_n) for which there exist numbers N and p (the period) such that $x_{n+p} = x_n$ for $n \geq N$ is almost convergent to the value $L(x_n) = \frac{1}{p}(x_N + x_{N+1} + \cdots + x_{N+p-1})$. For example, the periodic sequence $(1, 0, 0, 1, 0, 0, 1, \ldots)$ is almost convergent to $1/3$.

(iii) We say that a sequence (x_n) is *almost periodic* if for every $\varepsilon > 0$, there are two natural numbers N and r such that in every interval $(k, k + r)$, $k > 0$, at least one "ε-period" p exists. More precisely, $|x_{n+p} - x_n| < \varepsilon$ for $n \geq N$ must hold for this p. Thus, it is easy to see that every almost periodic sequence is almost convergent. But there are almost convergent sequences that are not almost periodic. For example, the sequence $x = (x_k)$ defined by

$$x_k = \begin{cases} 1 & \text{if } k = n^2, \\ 0 & \text{if } k \neq n^2, n \in \mathbb{N}, \end{cases}$$

is almost convergent (to 0) but not almost periodic.

Recently, Başar and Kirişçi [11] (see also [10]) defined the following space $\hat{f}$, which is linearly isomorphic to f:

$$\hat{f} := \left\{ x \in l_\infty : \lim_m \sum_{j=0}^{m} \frac{s x_{n-1+j} + r x_{n+j}}{m + 1} = \ell, \text{ uniformly in } n \right\},$$

where r and s are nonzero real numbers.

The notion of Banach limit and almost convergence was generalized by Duran [40] to $\mathcal{A}$-invariant mean and $\mathcal{A}$-almost convergence, respectively, and further studied by Mursaleen [91].

Let A be a positive regular matrix. It may be thought of as a bounded linear operator on l_∞. Let $\mathcal{A}$ be a semigroup of positive regular matrices A. Multiplication can be thought of either as matrix multiplication or as composition of operators on l_∞.

Definition 1.6 A continuous linear functional ϕ on l_∞ is said to be an $\mathcal{A}$-*invariant mean* if it has the following properties:

(i) $\phi(x) \geq 0$ if $x \geq 0$,
(ii) $\phi(e) = 1$, and
(iii) $\phi(Ax) = \phi(x)$ for all $A \in \mathcal{A}$ and $x \in l_\infty$.

We denote by $L(\mathcal{A})$ the set of all $\mathcal{A}$-invariant means. $\mathcal{A}$ is said to be *admissible* if $L(\mathcal{A})$ is nonempty.

Definition 1.7 Let $\mathcal{A}$ be an admissible semigroup of positive regular matrices. A bounded sequence $x = (x_k)$ is said to be $\mathcal{A}$-*almost convergent* to the number ℓ if $\phi(x) = \ell$ for all $\phi \in L(\mathcal{A})$, and we write this as $F(\mathcal{A})$-$\lim x = \ell$ or simply F-$\lim x = \ell$ when the semigroup is clear from the context. The space of all $\mathcal{A}$-almost convergent sequences is denoted by $F(\mathcal{A})$ or simply F.

Remark 1.8

(i) If $\mathcal{A}$ consists of the iterates of the shift matrix S defined by $(Sx)_n = x_{n+1}$, then the $\mathcal{A}$-invariant mean is reduced to the Banach limit, and $\mathcal{A}$-almost convergence is reduced to almost convergence due to Lorentz [63].
(ii) If $\mathcal{A}$ consists of the iterates of the operator T defined on l_∞ by $Tx = (x_{\sigma(n)})$, where σ is an injection of the set of positive integers into itself having no finite orbits, then the $\mathcal{A}$-invariant mean is reduced to the σ-mean, $\mathcal{A}$-almost convergence is reduced to σ-convergence due to Raimi [109] and Schaefer [114], and the space F is reduced to the space V_σ of σ-convergent sequences.

Remark 1.9   •

(i) If $x \in l_\infty$ is A-summable by some matrix $A \in \mathcal{A}$, say $\lim Ax = \ell$, then $F(\mathcal{A})$-$\lim x = \ell$, i.e., $c_A \subset F(\mathcal{A})$ with limit preserved.
(ii) It is clear that F is a direct sum of F_0 and $\mathbb{R}e$, where $\mathbb{R}$ is the set of real numbers, and F_0 consists of sequences that are $\mathcal{A}$-almost convergent to 0. F_0 is a closed subspace of l_∞ since it is an intersection of kernels of continuous linear functionals. Hence, F is closed as well. Note that both F and F_0 are invariant under $\mathcal{A}$.

1.2.2 Properties of the Method f

(1) Note that $c \subset f$ and for $x \in c$, $L(x) = \lim x$. That is, every convergent sequence is almost convergent to the same limit but not conversely. For example,

the sequence $x = (x_k)$ defined by

$$x_k = \begin{cases} 1 & \text{if } k \text{ is odd,} \\ 0 & \text{if } k \text{ is even,} \end{cases}$$

is not convergent, but it is almost convergent to $1/2$.

(2) f is a nonseparable closed subspace of $(l_\infty, \|\cdot\|_\infty)$.

(3) f is a BK-space with $\|\cdot\|_\infty$.

(4) f is nowhere dense in l_∞, dense in itself and closed, and therefore perfect.

(5) The method is not strong in spite of the fact that it contains certain classes of matrix methods for bounded sequences.

(6) Most of the commonly used matrix methods contain the method f, e.g., every almost convergent sequence is also (C, α) and (E, α)-summable ($\alpha > 0$) to its f-limit.

(7) The method f is equivalent to none of the matrix methods, i.e., the method f cannot be expressed in the form of a matrix method.

(8) According to (1.3), the method f seems to be related to the Cesàro method $(C, 1)$. In fact, the method $(C, 1)$ can be replaced in this definition by any other regular matrix method A satisfying certain conditions.

(9) Since $c \subset f \subset l_\infty$, we have $l_1 = l_\infty^\dagger \subset f^\dagger \subset c^\dagger = l_1$. That is, the $\dagger$-dual of f is l_1, where $\dagger$ stands for α, β, and γ.

Definition 1.10 Let $A = (a_{nk})_{n;k=0}^\infty$ be a regular matrix method, i.e., A transforms convergent sequences into convergent sequences leaving the limit invariant. A bounded sequence $x = (x_k)$ is said to be F_A-*summable* to the value ℓ if $y_{np} = \sum_k a_{nk} x_{k+p} \to \ell$ as $n \to \infty$ uniformly in $p = 0, 1, 2, \ldots$.

Note that if A is replaced by the $(C, 1)$ matrix, then the F_A-summability is reduced to the almost convergence.

We have the following result.

Theorem 1.11 *If the method A is regular, then every F_A-summable sequence is also almost convergent.*

1.3 Absolute and Strong Almost Convergence

It is natural to define the absolute and strong analogues of almost convergence in the same way as in the case of ordinary convergence (cf. [28–33, 64, 65, 87]).

Given an infinite series $\sum a_n$, which we will denote by a, let $x_n = a_0 + a_1 + \cdots + a_n$. Where limits are not stated, sums throughout are to be taken from 0 to ∞. We now extend the definition of $t_{mn}(x)$ to $m = -1$ by taking $t_{-1,n} = t_{-1,n}(x) = x_{n-1}$. A straightforward calculation shows that

$$\phi_{mn} = \phi_{mn}(a) = t_{mn} - t_{m-1,n} = \begin{cases} \frac{1}{m(m+1)} \sum_{v=1}^m a_{n+v} & (m \geq 1), \\ a_n & (m = 0). \end{cases} \tag{1.5}$$

Definition 1.12 We say that the series a (or the sequence x) is *absolutely almost convergent* if

$$\sum_m |\phi_{mn}| < \infty \quad \text{uniformly in } n, \tag{1.6}$$

and we denote the set of all absolutely almost convergent sequences by $\hat{l}$. Further, $\hat{l}$ is extended to $\hat{l}(p)$ in the same way as l was extended to $l(p)$ (see Simon [117]).

Let $p = (p_m)$ be a bounded sequence of positive numbers. We write

$$\psi_n(a) = \sum_m |\phi_{mn}|^{p_m} \tag{1.7}$$

whenever the series on the right converges. We write $\hat{l}(p)$ as the set of series for which (1.7) converges uniformly in n. If p_m is constant (say p), then $\hat{l}(p) = \hat{l}_p$, and we omit the suffix if $p = 1$. $\hat{l}(p)$ is a complete paranormed space, paranormed by $g_p(a) = \sup_n (\psi_n(a))^{1/H}$, where $H = \max(1, \sup_m p_m)$.

We may remark that the requirement in (1.6) and (1.7) a fortiori presupposes that the series converges for each n. In fact, we have the result, which asserts that if the series converges for one n, then it converges for any other value of n.

We note here some interesting properties of $\hat{l}_p$.

Theorem 1.13 *We have the following:*

(i) c and $\hat{l}$ *are not comparable.*
(ii) $\hat{l}$ *is not closed in* l_∞.
(iii) *It is possible for a series a to belong to $\hat{l}$ and hence to $\hat{l}_p$ for $p > 1$ without it being true that $a_n \to 0$ as $n \to \infty$.*
(iv) *If $p \leq \frac{1}{2}$, then $a \in \hat{l}_p$ implies that $a_n = 0$ $(n \geq 1)$.*
(v) *If $p \geq 1$, then $l_p \subset \hat{l}_p$, and this inclusion is proper. Further, if $a \in l_p$, then $\|a\|_{\hat{p}} \leq \|a\|_p$.*
(vi) *If $p < 1$, it is false that $l_p \subset \hat{l}_p$.*
(vii) *If $p > \frac{1}{2}$, the converse inclusion $\hat{l}_p \subset l_p$ is false.*
(viii) *For $1 \leq p < \infty$, $\hat{l}_p$ does not have AD.*
(ix) *For $1 < p < \infty$, $\hat{l}_p$ is not separable (and hence not reflexive). In particular, it is not a Hilbert space.*

Theorem 1.14 *Let p_m be a constant $p > \frac{1}{2}$. If the series $\sum_m |\phi_{mn}|^{p_m}$ converges for one value of n, then it converges for any other value of n. In fact,*

$$\sum_m |\phi_{mn}|^{p_m} = o(n^p), \tag{1.8}$$

and the result is best possible in the sense that even if we replace o by O, we cannot replace n^p by any function increasing less rapidly.

It is natural to consider the situation where x_{n_i+k} inside the summation in (1.2) is replaced by its modulus. That is, we consider

$$P(x) = \inf_{n_1, n_2, \ldots, n_p} \limsup_{k \to \infty} \frac{1}{p} \sum_{i=1}^{p} |x_{n_i+k}|, \tag{1.9}$$

which leads to the following concept (cf. [64]).

Definition 1.15 A sequence $x = (x_k)$ is said to be *strongly almost convergent* to the number ℓ if and only if

$$\lim_{m} \frac{1}{m+1} \sum_{i=0}^{m} |x_{n+i} - \ell| = 0 \quad \text{uniformly in } n. \tag{1.10}$$

Then we write $\ell = [f]\text{-}\lim x$, and by $[f]$ we denote the set of all strongly almost convergent sequences. We have $c \subset f \subset [f] \subset l_\infty$.

For more general cases, see (Mursaleen [84, 86]).

The following theorem gives the relation between absolute almost convergence and strong almost convergence (see Das and Mishra [29]).

Theorem 1.16 $\hat{l} \subset [f]$, and if $x \in [f]$, then $[f]\text{-}\lim x = f\text{-}\lim x = \ell$.

1.4 Statistical Convergence

In 1951, Fast [43] and Steinhaus [120] independently introduced an extension of the usual concept of sequential limit, which they called statistical convergence. In most convergence theories, it is desirable to have a criterion that can be used to verify the convergence without using the value of the limit. For this purpose, an analogue of the Cauchy convergence criterion is studied.

Definition 1.17 Let $K \subseteq \mathbb{N}$. Then the *natural density* of K is defined by

$$\delta(K) = \lim_{n} \frac{1}{n} |\{k \leq n; \ k \in K\}|,$$

where $|\{k \leq n; \ k \in K\}|$ denotes the number of elements of K not exceeding n.

For example, the set of even integers has natural density $\frac{1}{2}$, and the set of primes has natural density zero.

The number sequence x is said to be *statistically convergent* to the number L if for each $\epsilon > 0$,

$$\delta(K) = \lim_{n} \frac{1}{n} |\{k \leq n; \ |x_k - L| \geq \epsilon\}| = 0,$$

i.e.,

$$|x_k - L| < \epsilon \quad \text{a. a. } k. \tag{1.11}$$

In this case, we write $st\text{-}\lim x_k = L$. By the symbol st we denote the set of all statistically convergent sequences, and by st_0 the set of all statistically null sequences.

Note that every convergent sequence is statistically convergent to the same number, so that statistical convergence is a natural generalization of the usual convergence of sequences.

The sequence that converges statistically need not be convergent and also need not be bounded.

Example 1.18 Let $x = (x_k)$ be defined by

$$x_k = \begin{cases} k & \text{if } k \text{ is a square,} \\ 0 & \text{otherwise.} \end{cases}$$

Then $|\{k \le n : x_k \ne 0\}| \le \sqrt{n}$. Therefore, $st\text{-}\lim x_k \stackrel{\bullet}{=} 0$. Note that we could have assigned any values to x_k when k is a square, and we could still have $st\text{-}\lim x_k = 0$. But x is neither convergent nor bounded.

It is clear that if the inequality in (1.11) holds for all but finitely many k, then $\lim x_k = L$. It follows that $\lim x_k = L$ implies $st\text{-}\lim x_k = L$, so that statistical convergence may be considered as a regular summability method. This was observed by Schoenberg [115] along with the fact that the statistical limit is a linear functional on some sequence space. Salat [112] proved that the set of bounded statistically convergent (real) sequences is a closed subspace of the space of bounded sequences.

Definition 1.19 The number sequence x is said to be *statistically Cauchy sequence* if for every $\epsilon > 0$, there exists a number $N(=N(\epsilon))$ such that

$$|x_k - x_N| < \epsilon \quad \text{a. a. } k, \tag{1.12}$$

i.e.,

$$\lim_n \frac{1}{n} \left| \{k \le n : |x_n - x_N| \ge \epsilon\} \right| = 0.$$

In order to prove the equivalence of Definitions 1.17 and 1.19, we shall find it helpful to use a third (equivalent) one. This property states that for almost all k, the values x_k coincide with those of a convergent sequence.

Theorem 1.20 *The following statements are equivalent:*

(i) x *is a statistically convergent sequence.*
(ii) x *is a statistically Cauchy sequence.*

(iii) *x is a sequence for which there is a convergent sequence y such that $x_k = y_k$ a. a. k.*

Definition 1.21 Let $K = \{k_i\}$ be an index set, and let $\varphi^K = (\varphi_j^K)$ with

$$\varphi_j^K = \begin{cases} 1 & \text{if } j \in K, \\ 0 & \text{otherwise.} \end{cases}$$

For a nonnegative regular matrix A, if $A\varphi^K \in c$ (the space of convergent sequences), then $\delta_A(K) = \lim_n A_n \varphi^K$ is called the *A-density* [44] of K; thus,

$$\delta_A(K) = \lim_n \sum_{k \in K} a_{nk} = \lim_n \sum_i a_{n,k_i}.$$

A sequence $x = (x_k)$ is said to be *A-statistically convergent* to the number L if $\delta_A(K_\epsilon) = 0$ for every $\epsilon > 0$, where $K_\epsilon = \{k : |x_k - L| \geq \epsilon\}$. In this case, we write st_A-$\lim x_k = L$. By the symbol st_A we denote the set of all A-statistically convergent sequences, and by st_A^0 the set of all A-statistically null sequences [57].

1.5 Statistical Limit Points and Statistical Cluster Points

The number L is an ordinary limit point of a sequence x if there is a subsequence of x that converges to L; therefore, we define a statistical limit point by considering the density of such a subsequence [46].

Definition 1.22 The number λ is a *statistical limit point* of the number sequence x if there is a nonthin subsequence of x that converges to λ.

Notation For any number sequence x, let Λ_x denote the set of statistical limit points of x, and L_x denote the set of ordinary limit points of x.

Example 1.23 Let $x_k = 1$ if k is a square and $x_k = 0$ otherwise; then $L_x = \{0, 1\}$ and $\Lambda_x = \{0\}$. It is clear that $\Lambda_x \subseteq L_x$ for any sequence x. To show that Λ_x and L_x can be very different, we give a sequence x for which $\Lambda_x = \phi$ while $L_x = \mathbb{R}$, the set of real numbers.

Example 1.24 Let $(r_k)_{k=1}^\infty$ be a sequence whose range is the set of all rational numbers and define

$$x_k = \begin{cases} r_n & \text{if } k = n^2 \text{ for } n = 1, 2, 3, \ldots, \\ k & \text{otherwise.} \end{cases}$$

Since the set of squares has density zero, it follows that $\Lambda_x = \phi$, while the fact that $\{r_k : k \in \mathbb{N}\}$ is dense in $\mathbb{R}$ implies that $L_x = \mathbb{R}$.

A limit point L of a sequence x can be characterized by the statement "every open interval centered at L contains infinitely many terms of x." To form a statistical analogue of this criterion, we require the open interval to contain a nonthin subsequence, but we must avoid calling the center of the interval a statistical limit point for reasons that will be apparent shortly.

Definition 1.25 The number γ is a *statistical cluster point* of the number sequence x if for every $\epsilon > 0$, the set $\{k \in \mathbb{N} : |x_k - \gamma| < \epsilon\}$ does not have density zero. For a given sequence x, we let Γ_x denote the set of all statistical cluster points of x. It is clear that $\Gamma_x \subseteq L_x$ for every sequence x. The inclusion relationship between Γ_x and Λ_x is a bit more subtle.

Proposition 1.26 *For any number sequence x, $\Lambda_x \subseteq \Gamma_x$.*

Our experience with ordinary limit points may lead us to expect that Λ_x and Γ_x are equivalent. The next example shows that this is not always the case.

Example 1.27 Define the sequence x by

$$x_k = 1/p, \quad \text{where } k = 2^{p-1}(2q+1);$$

i.e., $p - 1$ is the number of factors of 2 in the prime factorization of k. It is easy to see that for each p, $\delta\{k : x_k = 1/p\} = 2^{-p} > 0$, whence $1/p \in \Lambda_x$. Also, $\delta\{k : 0 < x_k < 1/p\} = 2^{-p}$, so $0 \in \Gamma_x$, and we have $\Gamma_x = \{0\} \cup \{1/p\}_{p=1}^{\infty}$. Now we assert that $0 \notin \Lambda_x$ since if $(x)_K$ is a subsequence that has limit zero, then we can show that $\delta(K) = 0$. This is done by observing that for each p,

$$|K_n| = \left|\{k \in K_n : x_k \geq 1/p\}\right| + \left|\{k \in K_n : x_k < 1/p\}\right|$$

$$\leq O(1) + \left|\{k \in \mathbb{N} : x_k < 1/p\}\right| \leq O(1) + n/2^p.$$

Thus $\delta(K) \leq 2^{-p}$, and since p is arbitrary, this implies that $\delta(K) = 0$.

It is easy to prove that if x is a statistically convergent sequence, say $st\text{-}\lim x = \lambda$, then Λ_x and Γ_x are both equal to the singleton set $\{\lambda\}$. The converse is not true, as one can see by taking $x_k = [1+(-1)^k]k$. The following example presents a sequence x for which Γ_x is an interval while $\Lambda_x = \phi$.

Example 1.28 Let x be the sequence $\{0, 0, 1, 0, \frac{1}{2}, 1, 0, \frac{1}{3}, \frac{2}{3}, 1, \dots\}$. This sequence is uniformly distributed in $[0, 1]$, so we have that not only $L_x = [0, 1]$ but also the density of the x_ks in any subinterval of length d is d itself. Therefore, for any γ in $[0, 1]$,

$$\delta\left(\{k \in \mathbb{N} : x_k \in (\gamma - \epsilon, \gamma + \epsilon)\}\right) \geq \epsilon > 0.$$

Hence, $\Gamma_x = [0, 1]$. On the other hand, if $\lambda \in [0, 1]$ and $(x)_K$ is a subsequence that converges to λ, then we claim that $\delta\{K\} = 0$. To prove this assertion, let $\epsilon > 0$ be given and note that for each n,

$$|K_n| \leq \left|\{k \in K_n : |x_k - \lambda| < \epsilon\}\right| + \left|\{k \in K_n : |x_k - \lambda| \geq \epsilon\}\right|$$
$$\leq 2\epsilon n + O(1).$$

Consequently, $\delta\{k(j)\} \leq 2\epsilon$, and since ϵ is arbitrary, we conclude that $\delta\{k(j)\} = 0$. Hence, $\Lambda_x = \phi$.

1.6 Statistical Limit Superior and Statistical Limit Inferior

Throughout the chapter, k and n will always denote positive integers; x, y, and z will denote real number sequences; and $\mathbb{N}$ and $\mathbb{R}$ will denote the sets of positive integers and real numbers, respectively. If $K \subseteq \mathbb{N}$, then $K_n = \{k : k \leq n\}$, and $|K_n|$ denotes the cardinality of K_n.

For a real number sequence x, denote

$$B_x = \{b \in \mathbb{R} : \delta\{k : x_k > b\} \neq 0\};$$

similarly,

$$A_x = \{a \in \mathbb{R} : \delta\{k : x_k < a\} \neq 0\}.$$

Note that, throughout this chapter, the statement $\delta\{K\} \neq 0$ means that either $\delta\{K\} > 0$ or K does not have natural density.

Definition 1.29 If x is a real number sequence, then the *statistical limit superior* of x is given by

$$st\text{-}\limsup x = \begin{cases} \sup B_x & \text{if } B_x \neq \phi, \\ -\infty & \text{if } B_x = \phi. \end{cases}$$

Also, the *statistical limit inferior* of x is given by

$$st\text{-}\liminf x = \begin{cases} \inf A_x & \text{if } A_x \neq \phi, \\ +\infty & \text{if } A_x = \phi. \end{cases}$$

Example 1.30 A simple example will help to illustrate the concepts just defined. Let the sequence x be given by

$$x_k = \begin{cases} k & \text{if } k \text{ is an odd square,} \\ 2 & \text{if } k \text{ is an even square,} \\ 1 & \text{if } k \text{ is an odd nonsquare,} \\ 0 & \text{if } k \text{ is an even nonsquare.} \end{cases}$$

Note that although x is unbounded above, it is "statistically bounded" because the set of squares has density zero. Thus, $B_x = (-\infty, 1)$ and $st\text{-}\limsup x = 1$. Also,

x is not statistically convergent since it has two (disjoint) subsequences of positive density that converge to 0 and 1, respectively. Also note that the set of statistical cluster points of x is $\{0, 1\}$ and st-$\lim \sup x$ equals the greatest element while st-lim inf is the least element of this set. This observation suggests the main idea of the first theorem, which can be proved by a straightforward least upper bound argument [47].

1.7 Knopp Core and Statistical Core

Definition 1.31 The *core* or *K-core* of a real number sequence $x = (x_k)$ is defined to be the closed interval $[\lim \inf x, \lim \sup x]$. If x is a complex number sequence, then its core is defined as

$$K\text{-}core\{x\} = \bigcap_{n=1}^{\infty} C_n(x),$$

where $C_n(x)$ is the closed convex hull of $(x_k)_{k \geq n}$.

The well-known Knopp core theorem states as follows (see Knopp [56]).

Theorem 1.32 (Knopp's core theorem) *In order that* $L(Ax) \leq L(x)$ *for every bounded sequence* $x = (x_k)$, *it is necessary and sufficient that* $A = (a_{nk})$ *should be regular and* $\lim_n \sum_{k=0}^{\infty} |a_{nk}| = 1$, *where* $L(x) = \lim \sup x$.

Definition 1.33 If x is a statistically bounded sequence, then the *statistical core* of x is the closed interval $[st$-$\lim \inf x, st$-$\lim \sup x]$. In case x is not statistically bounded, st-$core\{x\}$ is defined accordingly as either $[st$-$\lim \inf x, \infty)$, $(-\infty, \infty)$, or $(-\infty, st$-$\lim \sup x]$. We shall denote the statistical core of x by st-$core\{x\}$. It is clear that for any real sequence x,

$$st\text{-}core\{x\} \subseteq K\text{-}core\{x\}.$$

1.8 Matrix Transformations

Definition 1.34 Let X and Y be two sequence spaces, and $A = (a_{nk})_{n;k=1}^{\infty}$ be an infinite matrix of real or complex numbers. We write $Ax = (A_n(x))$, $A_n(x) = \sum_k a_{nk} x_k$ if the series on the right converges for each n. If $x = (x_k) \in X$ implies that $Ax \in Y$, then we say that A defines a matrix transformation from X into Y, and we denote the class of such matrices by (X, Y).

Definition 1.35 A matrix $A = (a_{nk})_{n;k=1}^{\infty}$ is said be *conservative* if $Ax \in c$ for all $x \in c$. In addition, if $\lim Ax = \lim x$ for all $x \in c$, then A is said to be *regular*.

The well-known Silverman–Toeplitz conditions for A to be regular are as follows.

Theorem 1.36

(a) $A = (a_{nk})_{n;k=1}^{\infty}$ *is conservative if and only if*

    (i) $\|A\| = \sup_n \sum_k |a_{nk}| < \infty$,
    (ii) $\lim_{n\to\infty} a_{nk} = \alpha_k \in \mathbb{C}$ *for each* k, *and*
    (iii) $\lim_{n\to\infty} \sum_k a_{nk} = \alpha \in \mathbb{C}$.

(b) *A is regular if and only if conditions* (i), (ii) *with* $\alpha_k = 0$ *for each* k *and* (iii) *with* $\alpha = 1$ *hold.*

Lorentz defined and characterized the following class.

Definition 1.37 *A is said to be* strongly regular *(cf. [63]) if it sums all almost convergent sequences and* $\lim Ax = f\text{-}\lim x$ *for all* $x \in f$.

Theorem 1.38 *A is strongly regular if and only if*

(i) *A is regular, and*
(ii) $\lim_{n\to\infty} \sum_k |a_{nk} - a_{n,k+1}| = 0$.

King [55] used the idea of almost convergence to study the almost conservative and almost regular matrices.

Definition 1.39 A matrix $A = (a_{nk})_{n;k=1}^{\infty}$ is said be *almost conservative* if $Ax \in f$ for all $x \in c$. In addition, if $f\text{-}\lim Ax = \lim x$ for all $x \in c$, then A is said to be *almost regular*.

Theorem 1.40

(a) $A = (a_{nk})_{n;k=1}^{\infty}$ *is almost conservative if and only if*

    (i) $\|A\| < \infty$,
    (ii) $\lim_{p\to\infty} t(n, k, p) = \alpha_k \in \mathbb{C}$ *for each* k, *uniformly in* n, *and*
    (iii) $\lim_{p\to\infty} \sum_k t(n, k, p) = \alpha \in \mathbb{C}$, *uniformly in* n, *where*

$$t(n, k, p) = \frac{1}{p+1} \sum_{j=n}^{n+p} a_{jk}.$$

(b) *A is almost regular if and only if conditions* (i), (ii) *with* $\alpha_k = 0$ *for each* k *and* (iii) *with* $\alpha = 1$ *hold.*

Remark 1.41 Every regular matrix is almost regular (since $c \subset f$), but an almost regular matrix need not be regular. Let $C = (c_{nk})$ be defined by

$$c_{nk} = \begin{cases} \frac{1}{n+1}[1 + (-1)^n], & 0 \le k \le n, \\ 0, & n < k. \end{cases}$$

Then it is easy to see that the method C is almost regular but not regular (since $\lim_n \sum_k c_{nk}$ does not exist).

Eizen and Laush [42] considered the class of almost coercive matrices.

Definition 1.42 A matrix $A = (a_{nk})_{n;k=1}^{\infty}$ is said be *almost coercive* if $Ax \in f$ for all $x \in l_{\infty}$.

Theorem 1.43 $A = (a_{nk})_{n;k=1}^{\infty}$ *is almost coercive if and only if*

(i) $\|A\| < \infty$,
(ii) $\lim_{p \to \infty} t(n, k, p) = \alpha_k \in \mathbb{C}$ *for each k, uniformly in n, and*
(iii) $\lim_{p \to \infty} \sum_k |t(n, k, p) - \alpha_k| = 0$, *uniformly in n.*

Duran [39] considered the class of almost strongly regular matrices.

Definition 1.44 A matrix $A = (a_{nk})_{n;k=1}^{\infty}$ is said be *almost strongly regular* if $Ax \in f$ for all $x \in f$.

Theorem 1.45 $A = (a_{nk})_{n;k=1}^{\infty}$ *is almost strongly regular if and only if*

(i) *A is almost regular, and*
(ii) $\lim_{p \to \infty} \sum_k |t(n, k, p) - t(n, k+1, p)| = 0$, *uniformly in n.*

Theorem 1.46 *A complex matrix $A \in ([f], c)$ with $\lim Ax = [f]$-$\lim x$ if and only if A is regular and*

$$\sum_{k \in E} |\Delta a_{nk}| \to 0 \quad (n \to \infty)$$

for each set E that is uniformly of zero density, where $\Delta a_{nk} = a_{nk} - a_{n,k+1}$.

1.9 Exercises

1 Show that F is a BK-space with $\|.\|_{\infty}$.

2 Prove that F is a nonseparable closed subspace of $(l_{\infty}, \|.\|_{\infty})$.

3 Find α-, β-, and γ-duals of the sequence space $\hat{f}$.

4 Show that $\hat{l}(p)$ is a complete paranormed space, paranormed by $g_p(a) = \sup_n (\psi_n(a))^{1/H}$, where $H = \max(1, \sup_m p_m)$, and $\psi_n(a)$ is defined by (1.7).

5 Prove Theorem 1.13.

6   Give an example of a sequence that is (i) almost convergent but not statistically convergent, (ii) statistically convergent but not almost convergent.

7   Define the Banach core (B-core) of a real bounded sequence by using the sublinear functional $p(x)$ defined by (1.2) and establish its relation with K-core and st-core.

8   Define and characterize the σ-conservative, σ-regular, and σ-coercive matrices by using the notion of σ-convergence as described in Remark 1.8(ii).

Chapter 2
Almost Convergence of Double Sequences

The notion of almost convergence for ordinary (single) sequences was given by Lorentz [63], and for double sequences by Moricz and Rhoades [83]. In this chapter, we discuss the notion of almost convergence and almost Cauchy for double sequences. Some more related spaces for double sequences, associated sublinear functionals, and various inclusion relations are also studied.

2.1 Introduction

A double sequence $x = (x_{jk})$ of real or complex numbers is said to be *bounded* if $\|x\|_\infty = \sup_{j,k} |x_{jk}| < \infty$. The space of all bounded double sequences is denoted by $\mathcal{M}_u$.

A double sequence $x = (x_{jk})$ is said to *converge to the limit L in Pringsheim's sense* (shortly, *P-convergent to L*) [108] if for every $\varepsilon > 0$, there exists an integer N such that $|x_{jk} - L| < \varepsilon$ whenever $j, k > N$. In this case L is called the P-limit of x. If, in addition, $x \in \mathcal{M}_u$, then x is said to be *boundedly convergent to L in Pringsheim's sense* (shortly, *BP-convergent to L*). The sets of P-convergent and BP-convergent double sequences $x = (x_{jk})$ will be denoted by $\mathcal{C}_p$ and $\mathcal{C}_{bp}$, respectively.

A double sequence $x = (x_{jk})$ is said to *converge regularly to L* (shortly, *R-convergent to L*) if $x \in \mathcal{C}_p$ and the limits $x^j := \lim_k x_{jk}$ $(j \in \mathbb{N})$ and $x^k := \lim_j x_{jk}$ $(k \in \mathbb{N})$ exist. Note that, in this case, the limits $\lim_j \lim_k x_{jk}$ and $\lim_k \lim_j x_{jk}$ exist and are equal to the P-limit of x.

In general, for any notion of convergence ν, the space of all ν-convergent double sequences will be denoted by $\mathcal{C}_\nu$, the space of all ν-convergent to 0 double sequences by $\mathcal{C}_{\nu 0}$, and the limit of a ν-convergent double sequence x by $\nu\text{-}\lim_{j,k} x_{jk}$, where $\nu \in \{P, BP, R\}$.

The idea of almost convergence for single sequences was introduced by Lorentz [63], and for double sequences by Moricz and Rhoades [83].

M. Mursaleen, S.A. Mohiuddine, *Convergence Methods for Double Sequences and Applications*, DOI 10.1007/978-81-322-1611-7_2, © Springer India 2014

A double sequence $x = (x_{jk})$ of real numbers is said to be *almost convergent* to a limit L if

$$\lim_{\substack{p,q \to \infty \\ m,n > 0}} \sup \left| \frac{1}{pq} \sum_{j=m}^{m+p-1} \sum_{k=n}^{n+q-1} x_{jk} - L \right| = 0. \tag{2.1}$$

In this case, L is called the $\mathcal{F}$-limit of x, and we shall denote by $\mathcal{F}$ the space of all almost convergent double sequences, i.e.,

$$\mathcal{F} = \left\{ x = (x_{jk}) : \lim_{p,q \to \infty} |\tau_{pqst}(x) - L| = 0, \text{ uniformly in } s, t \right\},$$

where

$$\tau_{pqst}(x) = \frac{1}{(p+1)(q+1)} \sum_{j=0}^{p} \sum_{k=0}^{q} x_{j+s,k+t}.$$

Note that throughout the book, the notation lim for double sequences will represent P-lim.

If $m = n = 1$ in (2.1), then we get the $(C, 1, 1)$-convergence, and in this case, we write $x_{jk} \to \ell(C, 1, 1)$, where $\ell = (C, 1, 1)$-$\lim x$.

Recently, the concept of Banach limits for double sequences was defined in [101] as follows.

A linear functional $\mathcal{L}$ on $\mathcal{M}_u$ is said to be a *Banach limit* if it has the following properties:

(i) $\mathcal{L}(x) \geq 0$ if $x \geq 0$ (i.e., $x_{jk} \geq 0$ for all j, k),
(ii) $\mathcal{L}(\mathbf{e}) = 1$, where $\mathbf{e} = (e_{jk})$ with $e_{jk} = 1$ for all j, k, and
(iii) $\mathcal{L}(S_{11}x) = \mathcal{L}(x) = \mathcal{L}(S_{10}x) = \mathcal{L}(S_{01}x)$,

where the shift operators S_{01}, S_{10}, and S_{11} are defined by

$$S_{01}x = (x_{j,k+1}), \qquad S_{10}x = (x_{j+1,k}), \qquad S_{11}x = (x_{j+1,k+1}).$$

Let $\mathcal{B}_2$ be the set of all Banach limits on $\mathcal{M}_u$. A double sequence $x = (x_{jk})$ is said to be *almost convergent* to a number L if $\mathcal{L}(x) = L$ for all $\mathcal{L} \in \mathcal{B}_2$.

As in case of single sequences [63], Theorem 2.5 gives the equivalence of these two definitions.

The idea of strong almost convergence for single sequences is due to Maddox [64], and for double sequences, to Başarir [12].

A double sequence $x = (x_{jk})$ is said to be *strongly almost convergent* to a number L if

$$P\text{-}\lim_{p,q \to \infty} \frac{1}{(p+1)(q+1)} \sum_{j=0}^{p} \sum_{k=0}^{q} |x_{j+s,k+t} - L| = 0,$$

uniformly in s, t. By $[\mathcal{F}]$ we denote the space of all strongly almost convergent double sequences. Note that $[\mathcal{F}] \subset \mathcal{F} \subset \mathcal{M}_u$.

In [26], Čunjalo introduced the idea of almost Cauchy for double sequences. A double sequence $x = (x_{jk})$ is said to be *almost Cauchy* if for every $\epsilon > 0$, there exits $k \in \mathbb{N}$ such that

$$\left| \frac{1}{p_1 q_1} \sum_{j=0}^{p_1-1} \sum_{k=0}^{q_1-1} x_{j+s_1,k+t_1} - \frac{1}{p_2 q_2} \sum_{j=0}^{p_2-1} \sum_{k=0}^{q_2-1} x_{j+s_2,k+t_2} \right| < \epsilon$$

for all $p_1, p_2, q_1, q_2 > k$ and $(s_1, t_1), (s_2, t_2) \in \mathbb{N} \times \mathbb{N}$.

2.2 Some Auxiliary Results

Note that, in contrast to the single sequences, a P-convergent double sequence need not be almost convergent. However, every bounded convergent double sequence is almost convergent, and every almost convergent double sequence is also bounded, i.e., $\mathcal{C}_{BP} \subset \mathcal{F} \subset \mathcal{M}_u$, and each inclusion is proper.

We start with the following basic result.

Theorem 2.1 *Let a double sequence $x = (x_{jk})$ be BP-convergent to L. Then it is almost convergent to L, but the converse is not true in general.*

Proof For a given $\varepsilon > 0$, we choose $N, M \in \mathbb{N}$ with

$$|x_{jk} - L| < \frac{\varepsilon}{2} \quad (j, k \geq N), \tag{2.2}$$

and for N, there are $p_0, q_0 \in \mathbb{N}$ such that $p_0 > N$, $q_0 > M$, and

$$\frac{2(N+1)(M+1)\|x\|_\infty}{p_0 q_0} < \frac{\varepsilon}{2}. \tag{2.3}$$

Then, for $s, t \in \mathbb{N}$ and $p \geq p_0, q \geq q_0$, we obtain

$$\left| \frac{1}{pq} \sum_{j=s}^{s+p-1} \sum_{k=t}^{t+q-1} x_{jk} - L \right| \leq \frac{1}{pq} \sum_{j=s}^{s+p-1} \sum_{k=t}^{t+q-1} |x_{jk} - L|$$

$$\leq \begin{cases} \frac{1}{pq} pq \frac{\varepsilon}{2} & \text{if } s \geq N, \ t \geq M, \\ \frac{1}{pq} \sum_{j=s}^{N} \sum_{k=t}^{M} |x_{jk} - L| & \\ \quad + \frac{1}{pq} \sum_{j=N+1}^{s+p-1} \sum_{k=M+1}^{t+q-1} |x_{jk} - L| & \text{if } s < N, \ t < M, \end{cases}$$

$$\leq \begin{cases} \frac{\varepsilon}{2} & \text{if } s \geq N, \ t \geq M, \\ \frac{1}{pq}(N-s+1)(M-t+1)2\|x\|_\infty & \\ \quad + \frac{1}{pq}(s+p-1-N)(t+q-1-M)\frac{\varepsilon}{2} & \text{if } s < N, \ t < M, \end{cases}$$

$< \varepsilon$ $\big($by (2.2), (2.3) and the choice of p_0, q_0 and $N, M\big)$.

Thus, $x = (x_{jk})$ is almost convergent to L.

For converse, consider the double sequence $x = (x_{jk})$ defined by

$$
x_{jk} = \begin{cases} 1 & \text{if } j = k \text{ odd,} \\ -1 & \text{if } j = k \text{ even,} \\ 0 & \text{otherwise.} \end{cases}
$$

It is easy to see that $x = (x_{jk})$ is almost convergent to zero but not P-convergent, that is, $\mathcal{C}_{BP} \subsetneq \mathcal{F}$. $\square$

Theorem 2.2 *Every almost convergent double sequence $x = (x_{jk})$ is bounded, but the converse is not true in general.*

Proof Let $x = (x_{jk}) \in \mathcal{F}$ and $r_0, q_0 \in \mathbb{N}$ with

$$
\frac{1}{rq} \left| \sum_{j=s}^{s+r-1} \sum_{k=t}^{t+q-1} x_{jk} \right| \le 1 \quad (r \ge r_0, \ q \ge q_0, \text{ and } s, t \in \mathbb{N}) \tag{2.4}
$$

be given. Then, for $s, t \in \mathbb{N}$, we have

$$
|x_{st}| = \left| \sum_{j=s}^{s+r_0-1} \sum_{k=t}^{t+q_0-1} x_{jk} - \sum_{j=s+1}^{s+r_0-1} \sum_{k=t+1}^{t+q_0-1} x_{jk} \right|
$$

$$
= \left| \sum_{j=s}^{s+r_0-1} \sum_{k=t}^{t+q_0-1} (x_{jk} - L) - \sum_{j=s+1}^{s+r_0-1} \sum_{k=t+1}^{t+q_0-1} (x_{jk} - L) + L \right|
$$

$$
\le (r_0 + 1)(q_0 + 1) + r_0 q_0 + |L| \quad \text{by (2.4).}
$$

That is, $x = (x_{jk}) \in \mathcal{M}_u$.

For converse, let $x = (x_{jk})$ be defined as

$$
\begin{pmatrix}
1 & 0 & 0 & 1 & 1 & 1 & 1 & 0 & 0 & 0 & 0 & 0 & 0 & 0 & 0 & 1 & \cdot & \cdot & \cdot \\
1 & 0 & 0 & 1 & 1 & 1 & 1 & 0 & 0 & 0 & 0 & 0 & 0 & 0 & 0 & 1 & \cdot & \cdot & \cdot \\
1 & 0 & 0 & 1 & 1 & 1 & 1 & 0 & 0 & 0 & 0 & 0 & 0 & 0 & 0 & 1 & \cdot & \cdot & \cdot \\
\cdot & \cdot & \cdot & \cdot & \cdot & \cdot & \cdot & \cdot & \cdot & \cdot & \cdot & \cdot & \cdot & \cdot & \cdot & \cdot & \cdot & \cdot & \cdot \\
\cdot & \cdot & \cdot & \cdot & \cdot & \cdot & \cdot & \cdot & \cdot & \cdot & \cdot & \cdot & \cdot & \cdot & \cdot & \cdot & \cdot & \cdot & \cdot \\
\cdot & \cdot & \cdot & \cdot & \cdot & \cdot & \cdot & \cdot & \cdot & \cdot & \cdot & \cdot & \cdot & \cdot & \cdot & \cdot & \cdot & \cdot & \cdot
\end{pmatrix},
$$

i.e., in each row, there is one 1, then two 0s, then four 1s, then eight 0s, then sixteen 1s, etc.

Now, for n, m odd, the sum of the first $2^j 2^k$ elements will be at least $(2^{k-1} + 2^{k-3})2^j$, and so

$$\frac{1}{2^n 2^m} \sum_{j=1}^{n} \sum_{k=1}^{m} x_{jk} \geq \frac{2^n(2^{m-1} + 2^{m-3})}{2^n 2^m} = \frac{5}{8}.$$

Similarly, for n, m even,

$$\frac{1}{2^n 2^m} \sum_{j=1}^{n} \sum_{k=1}^{m} x_{jk} \leq \frac{2^n(2^{m-2} + 2^{m-3})}{2^n 2^m} = \frac{3}{8},$$

i.e., $x \notin \mathcal{F}$, but $x \in \mathcal{M}_u$. $\square$

Theorem 2.3 *Every double sequence $x = (x_{jk})$ is almost convergent if and only if it is almost Cauchy.*

Proof Let $x = (x_{jk})$ be almost convergent double sequence. Then, for every $\epsilon > 0$, there exists $k \in \mathbb{N}$ such that

$$\left| \frac{1}{pq} \sum_{j=0}^{p-1} \sum_{k=0}^{q-1} x_{j+s,k+t} - L \right| < \epsilon$$

for all $p, q > k$ and $(n, m) \in \mathbb{N} \times \mathbb{N}$. Therefore,

$$\left| \frac{1}{p_1 q_1} \sum_{j=0}^{p_1-1} \sum_{k=0}^{q_1-1} x_{j+s_1,k+t_1} - \frac{1}{p_2 q_2} \sum_{j=0}^{p_2-1} \sum_{k=0}^{q_2-1} x_{j+s_2,k+t_2} \right|$$

$$\leq \left| \frac{1}{p_1 q_1} \sum_{j=0}^{p_1-1} \sum_{k=0}^{q_1-1} x_{j+s_1,k+t_1} - L \right| + \left| \frac{1}{p_2 q_2} \sum_{j=0}^{p_2-1} \sum_{k=0}^{q_2-1} x_{j+s_2,k+t_2} - L \right|$$

$$< \frac{\epsilon}{2} + \frac{\epsilon}{2} = \epsilon$$

for all $p_1, p_2, q_1, q_2 > k$ and $(s_1, t_1), (s_2, t_2) \in \mathbb{N} \times \mathbb{N}$. Hence, $x = (x_{jk})$ is almost Cauchy.

Conversely, let $x = (x_{jk})$ be almost Cauchy. Then, for every $\epsilon > 0$, there exists $k \in \mathbb{N}$ such that

$$\left| \frac{1}{p_1 q_1} \sum_{j=0}^{p_1-1} \sum_{k=0}^{q_1-1} x_{j+s_1,k+t_1} - \frac{1}{p_2 q_2} \sum_{j=0}^{p_2-1} \sum_{k=0}^{q_2-1} x_{j+s_2,k+t_2} \right| < \frac{\epsilon}{2}$$

for all $p_1, p_2, q_1, q_2 > k$ and $(s_1, t_1), (s_2, t_2) \in \mathbb{N} \times \mathbb{N}$. Taking $s_1 = s_2 = s_0$ and $t_1 = t_2 = t_0$ in the above equation, we obtain that $\left(\frac{1}{pq} \sum_{j=0}^{p-1} \sum_{k=0}^{q-1} x_{j+s_0,k+t_0} \right)_{p,q=1}^{\infty}$ is a Cauchy sequence and hence convergent. Let

$$P\text{-}\lim_{p,q\to\infty}\frac{1}{pq}\sum_{j=0}^{p-1}\sum_{k=0}^{q-1}x_{j+s_0,k+t_0}=L.$$

Then, for every $\epsilon>0$, there exists $k_1\in\mathbb{N}$ such that

$$\left|\frac{1}{pq}\sum_{j=0}^{p-1}\sum_{k=0}^{q-1}x_{j+s_0,k+t_0}-L\right|<\frac{\epsilon}{2}$$

for all $p,q>k_1$. It follows that

$$\left|\frac{1}{pq}\sum_{j=0}^{p-1}\sum_{k=0}^{q-1}x_{j+s,k+t}-L\right|$$

$$\leq\left|\frac{1}{pq}\sum_{j=0}^{p-1}\sum_{k=0}^{q-1}x_{j+s,k+t}-\frac{1}{pq}\sum_{j=0}^{p-1}\sum_{k=0}^{q-1}x_{j+s_0,k+t_0}\right|+\left|\frac{1}{pq}\sum_{j=0}^{p-1}\sum_{k=0}^{q-1}x_{j+s_0,k+t_0}-L\right|$$

$$<\frac{\epsilon}{2}+\frac{\epsilon}{2}=\epsilon$$

for all $p,q>\max(k,k_1)$ and $(n,m)\in\mathbb{N}\times\mathbb{N}$. Hence, x is almost convergent to L. $\square$

2.3 Some Related Spaces of Double Sequences

In this section we introduce the following spaces involving the idea of Banach limit and almost convergence for double sequences. Such type of spaces for single sequences were studied by Das and Sahoo [31], and for double sequences by Mursaleen and Mohiuddine [100, 101].

Let $\mathbf{e}=(e_{jk})$ with $e_{jk}=1$ for all j,k. Then

$$w_2=\left\{x=(x_{jk}):\frac{1}{(m+1)(n+1)}\sum_{p=0}^{m}\sum_{q=0}^{n}\tau_{pqst}(x-L\mathbf{e})\longrightarrow 0\text{ as }m,n\longrightarrow\infty,\right.$$

$$\left.\text{uniformly in }s,t,\text{ for some }L\right\},$$

$$[w_2]=\left\{x=(x_{jk}):\frac{1}{(m+1)(n+1)}\sum_{p=0}^{m}\sum_{q=0}^{n}|\tau_{pqst}(x-L\mathbf{e})|\longrightarrow 0\text{ as }m,n\longrightarrow\infty,\right.$$

$$\left.\text{uniformly in }s,t,\text{ for some }L\right\},$$

$$[w]_2 = \left\{ x = (x_{jk}) : \frac{1}{(m+1)(n+1)} \sum_{p=0}^{m} \sum_{q=0}^{n} \tau_{pqst}\left(|x - Le|\right) \longrightarrow 0 \text{ as } m, n \longrightarrow \infty, \right.$$

$$\left. \text{uniformly in } s, t, \text{ for some } L \right\},$$

$$\hat{w}_2 = \left\{ x = (x_{jk}) : \sum_{p=0}^{\infty} \sum_{q=0}^{\infty} |d_{pqst} - d_{p-1,q,s,t} - d_{p,q-1,s,t} + d_{p-1,q-1,s,t}| \right.$$

$$\left. \text{converges uniformly in } s, t \right\},$$

$$\hat{\hat{w}}_2 = \left\{ x = (x_{jk}) : \right.$$

$$\left. \sup_{s,t} \sum_{p=0}^{\infty} \sum_{q=0}^{\infty} |d_{pqst} - d_{p-1,q,s,t} - d_{p,q-1,s,t} + d_{p-1,q-1,s,t}| < \infty \right\},$$

where

$$d_{m,n,s,t} = d_{m,n,s,t}(x) = \frac{1}{(m+1)(n+1)} \sum_{p=0}^{m} \sum_{q=0}^{n} \tau_{pqst}(x)$$

and

$$d_{0,0,s,t}(x) = \tau_{0,0,s,t} = x_{s,t}, \qquad d_{-1,0,s,t}(x) = \tau_{-1,0,s,t}(x) = x_{s-1,t},$$

$$d_{0,-1,s,t}(x) = \tau_{0,-1,s,t}(x) = x_{s,t-1}, \qquad d_{-1,-1,s,t}(x) = \tau_{-1,-1,s,t}(x) = x_{s-1,t-1}.$$

By $(C_2, 2)$, we denote the space of *Cesàro summable* double sequences of order 2, which is defined by

$$(C_2, 2) = \left\{ x = (x_{jk}) : \frac{1}{(m+1)(n+1)} \sum_{p=0}^{m} \sum_{q=0}^{n} \tau_{p,q,0,0}(x) \longrightarrow L \text{ as } m, n \longrightarrow \infty \right\},$$

and by $[C_2, 2]$, we denote the space of *strongly Cesàro summable* double sequences of order 2, which is defined by

$$[C_2, 2] = \left\{ x = (x_{jk}) : \frac{1}{(m+1)(n+1)} \sum_{p=0}^{m} \sum_{q=0}^{n} |\tau_{p,q,0,0}(x) - L| \longrightarrow 0 \right.$$

$$\left. \text{as } m, n \longrightarrow \infty \right\}.$$

Remark 2.4 If $[w_2]$-$\lim x = L$, that is,

$$\frac{1}{(m+1)(n+1)} \sum_{p=0}^{m} \sum_{q=0}^{n} |\tau_{pqst} - L| \longrightarrow 0$$

as $m, n \longrightarrow \infty$, uniformly in s, t, then

$$\frac{1}{(m+1)(n+1)} \sum_{p=0}^{m} \sum_{q=0}^{n} \left| \frac{1}{p+1} \sum_{j=0}^{p} \tau_{jqst} - L \right| \longrightarrow 0$$

and

$$\frac{1}{(m+1)(n+1)} \sum_{p=0}^{m} \sum_{q=0}^{n} \left| \frac{1}{q+1} \sum_{k=0}^{q} \tau_{pkst} - L \right| \longrightarrow 0.$$

2.4 Associated Sublinear Functionals

Let G be any sublinear functional on $\mathcal{M}_u$. We write $\{\mathcal{M}_u, G\}$ to denote the set of all linear functionals F on $\mathcal{M}_u$ such that $F \leq G$, i.e., $F(x) \leq G(x)$ for all $x = (x_{jk}) \in \mathcal{M}_u$.

Now we define the following functionals on the space $\mathcal{M}_u$ of real bounded double sequences:

$$\phi(x) = \limsup_{m,n} \sup_{s,t} \frac{1}{(m+1)(n+1)} \sum_{p=0}^{m} \sum_{q=0}^{n} \tau_{pqst}(x),$$

$$\psi(x) = \limsup_{m,n} \sup_{s,t} \frac{1}{(m+1)(n+1)} \sum_{p=0}^{m} \sum_{q=0}^{n} |\tau_{pqst}(x)|,$$

$$\theta(x) = \limsup_{m,n} \sup_{s,t} \frac{1}{(m+1)(n+1)} \sum_{p=0}^{m} \sum_{q=0}^{n} \tau_{pqst}(|x|),$$

$$\xi(x) = \limsup_{p,q} \sup_{s,t} \tau_{pqst}(x),$$

$$\eta(x) = \limsup_{p,q} \sup_{s,t} \tau_{pqst}(|x|),$$

where $|x| = (|x_{jk}|)_{j,k=1}^{\infty}$.

It can be easily verified that each of the above functionals is finite, well defined, and sublinear on $\mathcal{M}_u$.

A sublinear functional G is said to *generate* Banach limits if $F \in \{\mathcal{M}_u, G\}$ is a Banach limit, and it is said to *dominate* Banach limits if $F \in \mathcal{B}_2$ implies $F \in \{\mathcal{M}_u, G\}$.

In the following theorem, we characterize the space $\mathcal{M}_u \cap w_2$ in terms of the sublinear functional ϕ.

Theorem 2.5   *We have the following:*

(i) *The sublinear functional ϕ both dominates and generates Banach limits, i.e.,*
   $\phi(x) = \xi(x)$ *for all* $x = (x_{jk}) \in \mathcal{M}_u$.
(ii) $\mathcal{M}_u \cap w_2 = \{x = (x_{jk}) \in \mathcal{M}_u : \phi(x) = -\phi(-x)\} = \mathcal{F}$.

Proof (i) By the definition of ξ, for given $\epsilon > 0$, there exist p_0, q_0 such that

$$\tau_{pqst}(x) < \xi(x) + \epsilon$$

for $p > p_0, q > q_0$ and all s, t. This implies that

$$\phi(x) \le \xi(x) + \epsilon$$

for all $x = (x_{jk}) \in \mathcal{M}_u$. Since ϵ is arbitrary, we have $\phi(x) \le \xi(x)$ for all $x = (x_{jk}) \in \mathcal{M}_u$, and hence,

$$\{\mathcal{M}_u, \phi\} \subset \{\mathcal{M}_u, \xi\} = \mathcal{B}_2 \tag{2.5}$$

i.e., ϕ generates Banach limits.

Conversely, suppose that $\mathcal{L} \in \mathcal{B}_2$. Since $\mathcal{L}$ is shift invariant, i.e., $\mathcal{L}(S_{11}x) = \mathcal{L}(x) = \mathcal{L}(S_{10}x) = \mathcal{L}(S_{01}x)$, we have

$$\mathcal{L}(x) = \mathcal{L}\left(\frac{1}{(p+1)(q+1)} \sum_{j=0}^{p} \sum_{k=0}^{q} x_{j+s,k+t} \right)$$

$$= \mathcal{L}\left(\tau_{pqst}(x) \right)$$

$$= \mathcal{L}\left(\frac{1}{(m+1)(n+1)} \sum_{p=0}^{m} \sum_{q=0}^{n} \tau_{pqst}(x) \right). \tag{2.6}$$

But it follows from the definition of ϕ that for given $\epsilon > 0$, there exist m_0, n_0 such that

$$\frac{1}{(m+1)(n+1)} \sum_{p=0}^{m} \sum_{q=0}^{n} \tau_{pqst}(x) < \phi(x) + \epsilon \tag{2.7}$$

for $m > m_0, n > n_0$ and all s, t. Hence, by (2.7) and properties (i) and (ii) of Banach limits, we have

$$\mathcal{L}\left(\frac{1}{(m+1)(n+1)} \sum_{p=0}^{m} \sum_{q=0}^{n} \tau_{pqst}(x) \right) < \mathcal{L}((\phi(x) + \epsilon)e) = \phi(x) + \epsilon \tag{2.8}$$

for $m > m_0$, $n > n_0$ and all s, t, where $\mathbf{e}$ is defined at the beginning of Sect. 2.3. Since ϵ was arbitrary, it follows from (2.6) and (2.8) that

$$\mathcal{L}(x) \leq \phi(x) \quad \text{for all } x = (x_{jk}) \in \mathcal{M}_u.$$

Hence,

$$\mathcal{B}_2 \subset \{\mathcal{M}_u, \phi\}, \tag{2.9}$$

that is, ϕ dominates Banach limits.

Combining (2.5) and (2.9), we get

$$\{\mathcal{M}_u, \xi\} = \{\mathcal{M}_u, \phi\},$$

which implies that ϕ dominates and generates Banach limits and $\phi(x) = \xi(x)$ for all $x \in \mathcal{M}_u$.

(ii) As a consequence of the Hahn–Banach theorem, $\{\mathcal{M}_u, \phi\}$ is nonempty, and a linear functional $F \in \{\mathcal{M}_u, \phi\}$ is not necessarily uniquely defined at any particular value of x. This is evident in the manner the linear functionals are constructed. But in order that all the functionals $\{\mathcal{M}_u, \phi\}$ coincide at $x = (x_{jk})$, it is necessary and sufficient that

$$\phi(x) = -\phi(-x) \tag{2.10}$$

and

$$\limsup_{m,n} \sup_{s,t} \frac{1}{(m+1)(n+1)} \sum_{p=0}^{m} \sum_{q=0}^{n} \tau_{pqst}(x)$$

$$= \liminf_{m,n} \sup_{s,t} \frac{1}{(m+1)(n+1)} \sum_{p=0}^{m} \sum_{q=0}^{n} \tau_{pqst}(x). \tag{2.11}$$

Equality (2.11) holds if and only if

$$\frac{1}{(m+1)(n+1)} \sum_{p=0}^{m} \sum_{q=0}^{n} \tau_{pqst}(x) \longrightarrow L(\text{say}) \quad \text{as } m, n \longrightarrow \infty,$$

uniformly in s, t, and hence, $x = (x_{jk}) \in w_2 \cap \mathcal{M}_u$ for all $x = (x_{jk}) \in \mathcal{M}_u$, while Eq. (2.10) is equivalent to

$$\xi(x) = -\xi(-x),$$

which holds if and only if $x = (x_{jk}) \in \mathcal{F}$. $\qquad\square$

In the following theorem, we characterize the spaces $[w_2] \cap \mathcal{M}_u$ and $[w]_2 \cap \mathcal{M}_u$ in terms of the sublinear functionals.

Theorem 2.6 *We have the following*:

$$[w_2] \cap \mathcal{M}_u = \{x = (x_{jk}) : \psi(x - L\mathbf{e}) = 0, \text{ for some } L\}$$
$$= \{x = (x_{jk}) : F(x - L\mathbf{e}) = 0 \text{ for all } F \in \{\mathcal{M}_u, \psi\} \text{ for some } L\}.$$

Proof Without loss of generality, we assume that $L = 0$. Now, as in Theorem 2.5(ii), as a consequence of the Hahn–Banach theorem, $\{\mathcal{M}_u, \psi\}$ is nonempty, and a linear functional $F \in \{\mathcal{M}_u, \psi\}$ is not necessarily uniquely defined at any particular value of x. In order that all the functionals $\{\mathcal{M}_u, \psi\}$ coincide at $x = (x_{jk})$, it is necessary and sufficient that

$$\psi(x) = -\psi(-x).$$

Hence, in general, the construction of the sublinear functional ψ and the definition of $[w_2]$ together suggest that $x = (x_{jk}) \in [w_2] \cap \mathcal{M}_u$ if and only if

$$\psi(x - L\mathbf{e}) = -\psi(L\mathbf{e} - x). \tag{2.12}$$

Since $\psi(x) = \psi(-x)$, (2.12) reduces to

$$\psi(x - L\mathbf{e}) = 0. \tag{2.13}$$

Now, if $F \in \{\mathcal{M}_u, \psi\}$, then from (2.13) and from the linearity of F we have

$$F(x - L\mathbf{e}) = 0.$$

Conversely, suppose that $F(x - L\mathbf{e}) = 0$ for all $F \in \{\mathcal{M}_u, \psi\}$ and hence, by the Hahn–Banach theorem, there exists $F_0 \in \{\mathcal{M}_u, \psi\}$ such that $F_0(x) = \psi(x)$. Hence,

$$0 = F_0(x - L\mathbf{e}) = \psi(x - L\mathbf{e}).$$

This completes the proof of the theorem. $\qquad\qquad\qquad\qquad\qquad\qquad\qquad\square$

2.5 Some Basic Lemmas

In this section, we give some important lemmas, which will be used in the proofs of our main results.

Lemma 2.7 (Abel's transformation for double series)

$$\sum_{j=1}^{p}\sum_{k=1}^{q} v_{jk}(u_{jk} - u_{j+1,k} - u_{j,k+1} + u_{j+1,k+1})$$

$$= \sum_{j=1}^{p}\sum_{k=1}^{q} u_{jk}(\Delta_{11}v_{jk}) - \sum_{j=1}^{p} u_{j,q+1}(\Delta_{10}v_{jq}) - \sum_{k=1}^{q} u_{p+1,k}(\Delta_{01}v_{pk}) + u_{p+1,q+1}v_{pq},$$

where

$$\Delta_{10} v_{jq} = v_{jq} - v_{j-1,q}, \qquad \Delta_{01} v_{pk} = v_{pk} - v_{p,k-1} \quad and$$

$$\Delta_{11} v_{jk} = v_{jk} - v_{j-1,k} - v_{j,k-1} + v_{j-1,k-1}.$$

Proof Abel's transformation for single series is

$$\sum_{i=1}^{m} v_i(u_i \mp u_{i+1}) = \sum_{i=1}^{m} u_i(v_i \mp v_{i-1}) \mp u_{m+1} v_m. \tag{2.14}$$

Now we prove Abel's transformation for double series:

$$\sum_{j=1}^{p}\sum_{k=1}^{q} v_{jk}(u_{jk} - u_{j+1,k} - u_{j,k+1} + u_{j+1,k+1})$$

$$= \sum_{k=1}^{q}\left[\sum_{j=1}^{p} v_{jk}(u_{jk} - u_{j+1,k}) - \sum_{j=1}^{p} v_{jk}(u_{j,k+1} - u_{j+1,k+1})\right]$$

by using (2.14) we have

$$= \sum_{k=1}^{q}\left[\sum_{j=1}^{p} u_{jk}(v_{jk} - v_{j-1,k}) - u_{p+1,k} v_{pk}\right.$$

$$\left. - \sum_{j=1}^{p} u_{j,k+1}(v_{jk} - v_{j-1,k}) + u_{p+1,k+1} v_{pk}\right]$$

$$= \sum_{j=1}^{p}\left[\sum_{k=1}^{q} v_{jk}(u_{jk} - u_{j,k+1}) - \sum_{k=1}^{q} v_{j-1,k}(u_{jk} - u_{j,k+1})\right] - \sum_{k=1}^{q} u_{p+1,k} v_{pk}$$

$$+ \sum_{k=1}^{q} u_{p+1,k+1} v_{pk}$$

and now again using (2.14), we get

$$= \sum_{j=1}^{p}\left[\sum_{k=1}^{q} u_{jk}(v_{jk} - v_{j,k-1}) - u_{j,q+1} v_{jq} - \sum_{k=1}^{q} u_{jk}(v_{j-1,k} - v_{j-1,k-1})\right.$$

$$\left. + u_{j,q+1} v_{j-1,q}\right] - \sum_{k=1}^{q} u_{p+1,k} v_{pk} + \sum_{k=1}^{q} u_{p+1,k} v_{p,k-1} + u_{p+1,q+1} v_{pq}$$

$$= \sum_{j=1}^{p}\sum_{k=1}^{q} u_{jk}(v_{jk} - v_{j,k-1} - v_{j-1,k} + v_{j-1,k-1}) - \sum_{j=1}^{p} u_{j,q+1} v_{jq}$$

$$+ \sum_{j=1}^{p} u_{j,q+1} v_{j-1,q} - \sum_{k=1}^{q} u_{p+1,k} v_{pk} + \sum_{k=1}^{q} u_{p+1,k} v_{p,k-1} + u_{p+1,q+1} v_{pq}$$

$$= \sum_{j=1}^{p} \sum_{k=1}^{q} u_{jk} (\Delta_{11} v_{jk}) - \sum_{j=1}^{p} u_{j,q+1} (\Delta_{10} v_{jq}) - \sum_{k=1}^{q} u_{p+1,k} (\Delta_{01} v_{pk})$$

$$+ u_{p+1,q+1} v_{pq}. \qquad \qquad \square$$

Another form of Abel's transformation for double series is given by Altay and Başar [4].

Lemma 2.8 $[w_2]\text{-}\lim x = L$ *if and only if*

(i) $w_2\text{-}\lim x = L$;

(ii) $\frac{1}{uv} \sum_{m=1}^{u} \sum_{n=1}^{v} |T_1(m, n, s, t) - L| \longrightarrow 0 \ (u, v \longrightarrow \infty)$ *uniformly in* s, t;

(iii) $\frac{1}{uv} \sum_{m=1}^{u} \sum_{n=1}^{v} |T_2(m, n, s, t) - L| \longrightarrow 0 \ (u, v \longrightarrow \infty)$ *uniformly in* s, t;

(iv) $\frac{1}{uv} \sum_{m=1}^{u} \sum_{n=1}^{v} |\tau_{mnst} + d_{mnst} - T_1(m, n, s, t) - T_2(m, n, s, t)| \longrightarrow 0 \ (u, v \longrightarrow \infty)$ *uniformly in* s, t, *where*

$$T_1(m, n, s, t) = \frac{1}{(m+1)} \sum_{p=0}^{m} \tau_{pnst} \quad and \quad T_2(m, n, s, t) = \frac{1}{(n+1)} \sum_{q=0}^{n} \tau_{mqst}.$$

Proof Let $[w_2]\text{-}\lim x = L$. Then, obviously, $w_2\text{-}\lim x = L$. (ii) and (iii) follow immediately from Remark 2.4. Now we have

$$\frac{1}{uv} \sum_{m=1}^{u} \sum_{n=1}^{v} |\tau_{mnst} + d_{mnst} - T_1(m, n, s, t) - T_2(m, n, s, t)|$$

$$= \frac{1}{uv} \sum_{m=1}^{u} \sum_{n=1}^{v} |\tau_{mnst} - L + d_{mnst} - L - T_1(m, n, s, t) + L - T_2(m, n, s, t) + L|$$

$$\leq \frac{1}{uv} \sum_{m=1}^{u} \sum_{n=1}^{v} (|\tau_{mnst} - L| + |d_{mnst} - L| + |T_1(m, n, s, t) - L|$$

$$+ |T_2(m, n, s, t) - L|)$$

$\longrightarrow 0$ as $u, v \longrightarrow \infty$, uniformly in s, t, since

(a) $[w_2]\text{-}\lim x = L$ implies that the first sum tends to zero;

(b) (ii) and (iii) imply that the third and fourth sums tend to zero;

(c) (i) implies that $d_{mnst} \longrightarrow L \ (m, n \longrightarrow \infty)$ uniformly in s, t, and so the second sum tends to zero.

Conversely, suppose that the conditions hold. We have

$$\frac{1}{uv} \sum_{m=1}^{u} \sum_{n=1}^{v} |\tau_{mnst} - L|$$

$$\leq \frac{1}{uv} \sum_{m=1}^{u} \sum_{n=1}^{v} |\tau_{mnst} + d_{mnst} - T_1(m, n, s, t) - T_2(m, n, s, t)|$$

$$+ \frac{1}{uv} \sum_{m=1}^{u} \sum_{n=1}^{v} |d_{mnst} - L| + \frac{1}{uv} \sum_{m=1}^{u} \sum_{n=1}^{v} |T_1(m, n, s, t) - L|$$

$$+ \frac{1}{uv} \sum_{m=1}^{u} \sum_{n=1}^{v} |T_2(m, n, s, t) - L|$$

$\longrightarrow 0$ as $u, v \longrightarrow \infty$, uniformly in s, t.   □

Lemma 2.9 *We have*

$$\tau_{mnst} + d_{mnst} - T_1(m, n, s, t) - T_2(m, n, s, t)$$
$$= mn[d_{mnst} - d_{m-1,n,s,t} - d_{m,n-1,s,t} + d_{m-1,n-1,s,t}].$$

Proof We shall use the equality

$$d_{mnst} - d_{m-1,n,s,t} - d_{m,n-1,s,t} + d_{m-1,n-1,s,t}$$

$$= \left[\frac{1}{(m+1)(n+1)} \sum_{p=0}^{m} \sum_{q=0}^{n} \tau_{pqst} - \frac{1}{m(n+1)} \sum_{p=0}^{m-1} \sum_{q=0}^{n} \tau_{pqst} \right]$$

$$- \left[\frac{1}{(m+1)n} \sum_{p=0}^{m} \sum_{q=0}^{n-1} \tau_{pqst} - \frac{1}{mn} \sum_{p=0}^{m-1} \sum_{q=0}^{n-1} \tau_{pqst} \right]. \qquad (2.15)$$

First, we solve the expression in the first bracket:

$$\left[\frac{1}{(m+1)(n+1)} \sum_{p=0}^{m} \sum_{q=0}^{n} \tau_{pqst} - \frac{1}{m(n+1)} \sum_{p=0}^{m-1} \sum_{q=0}^{n} \tau_{pqst} \right]$$

$$= \frac{1}{m(m+1)(n+1)} \left[\sum_{q=0}^{n} \left(m \sum_{p=0}^{m} \tau_{pqst} - (m+1) \sum_{p=0}^{m-1} \tau_{pqst} \right) \right]$$

$$= \frac{1}{m(m+1)(n+1)} \sum_{q=0}^{n} \left[m\tau_{mqst} - \sum_{p=0}^{m-1} \tau_{pqst} \right]$$

$$= \frac{1}{m(m+1)(n+1)} \sum_{q=0}^{n} \left[(m+1)\tau_{mqst} - \sum_{p=0}^{m} \tau_{pqst} \right]$$

$$= \frac{1}{m(n+1)} \sum_{q=0}^{n} \left[\tau_{mqst} - \frac{1}{(m+1)} \sum_{p=0}^{m} \tau_{pqst} \right]$$

$$= \frac{1}{m(n+1)} \sum_{q=0}^{n} \tau_{mqst} - \frac{1}{m(m+1)(n+1)} \sum_{p=0}^{m} \sum_{q=0}^{n} \tau_{pqst}$$

$$= \frac{1}{m(n+1)} \sum_{q=0}^{n} \tau_{mqst} - \frac{1}{m} d_{mnst}. \tag{2.16}$$

For the expression in the second bracket, we have

$$\frac{1}{(m+1)n} \sum_{p=0}^{m} \sum_{q=0}^{n-1} \tau_{pqst} - \frac{1}{mn} \sum_{p=0}^{m-1} \sum_{q=0}^{n-1} \tau_{pqst}$$

$$= \frac{1}{n} \sum_{q=0}^{n-1} \left[\frac{1}{m(m+1)} \left\{ m \sum_{p=0}^{m} \tau_{pqst} - (m+1) \sum_{p=0}^{m-1} \tau_{pqst} \right\} \right]$$

$$= \frac{1}{n} \sum_{q=0}^{n-1} \left[\frac{1}{m(m+1)} \left\{ m\tau_{mqst} - \sum_{p=0}^{m-1} \tau_{pqst} \right\} \right]$$

$$= \frac{1}{n} \sum_{q=0}^{n-1} \left[\frac{1}{m(m+1)} \left\{ (m+1)\tau_{mqst} - \sum_{p=0}^{m} \tau_{pqst} \right\} \right]$$

$$= \frac{1}{mn} \sum_{q=0}^{n-1} \tau_{mqst} - \frac{1}{m(m+1)n} \sum_{p=0}^{m} \sum_{q=0}^{n-1} \tau_{pqst}$$

$$= \frac{1}{mn} \sum_{q=0}^{n-1} \tau_{mqst} - \frac{1}{m} d_{m,n-1,s,t}. \tag{2.17}$$

Substituting (2.16) and (2.17) into (2.15), we get

$$d_{mnst} - d_{m-1,n,s,t} - d_{m,n-1,s,t} + d_{m-1,n-1,s,t}$$

$$= \frac{1}{m(n+1)} \sum_{q=0}^{n} \tau_{mqst} - \frac{1}{mn} \sum_{q=0}^{n-1} \tau_{mqst} - \frac{1}{m} d_{mnst} + \frac{1}{m} d_{m,n-1,s,t}$$

$$= \frac{1}{mn(n+1)} \left[n \sum_{q=0}^{n} \tau_{mqst} - (n+1) \sum_{q=0}^{n-1} \tau_{mqst} \right] - \frac{1}{m} (d_{mnst} - d_{m,n-1,s,t})$$

$$= \frac{1}{mn(n+1)} \left[n\tau_{mnst} - \sum_{q=0}^{n-1} \tau_{mqst} \right] - \frac{1}{m}(d_{mnst} - d_{m,n-1,s,t})$$

$$= \frac{1}{mn(n+1)} \left[(n+1)\tau_{mnst} - \sum_{q=0}^{n} \tau_{mqst} \right] - \frac{1}{m}(d_{mnst} - d_{m,n-1,s,t})$$

$$= \frac{1}{mn} \left[\tau_{mnst} - \frac{1}{(n+1)} \sum_{q=0}^{n} \tau_{mqst} \right] - \frac{1}{m}(d_{mnst} - d_{m,n-1,s,t}). \qquad (2.18)$$

We know that

$$d_{mnst} = \frac{1}{(m+1)(n+1)} \sum_{p=0}^{m} \sum_{q=0}^{n} \tau_{pqst}$$

$$= \frac{1}{(m+1)(n+1)} \left[\sum_{p=0}^{m-1} \sum_{q=0}^{n} \tau_{pqst} + \sum_{q=0}^{n} \tau_{mqst} \right] \qquad (2.19)$$

and

$$d_{m-1,n,s,t} = \frac{1}{m(n+1)} \sum_{p=0}^{m-1} \sum_{q=0}^{n} \tau_{pqst}. \qquad (2.20)$$

From (2.19) and (2.20) we have

$$(m+1)d_{mnst} - md_{m-1,n,s,t} = \frac{1}{(n+1)} \sum_{q=0}^{n} \tau_{mqst}. \qquad (2.21)$$

Thus, (2.18) becomes

$$= \frac{1}{mn} \left[\tau_{mnst} - (m+1)d_{mnst} + md_{m-1,n,s,t} \right] - \frac{1}{m}(d_{mnst} - d_{m,n-1,s,t})$$

$$= \frac{1}{mn} \left[\tau_{mnst} - d_{mnst} - m(d_{mnst} - d_{m-1,n,s,t}) - n(d_{mnst} - d_{m,n-1,s,t}) \right]. \qquad (2.22)$$

Also, (2.21) can be written as

$$m(d_{mnst} - d_{m-1,n,s,t}) = \frac{1}{(n+1)} \sum_{q=0}^{n} \tau_{mqst} - d_{mnst}. \qquad (2.23)$$

Similarly, we can write

$$n(d_{mnst} - d_{m,n-1,s,t}) = \frac{1}{(m+1)} \sum_{q=0}^{m} \tau_{pnst} - d_{mnst}. \qquad (2.24)$$

Using (2.23) and (2.24) in (2.22), we get

$$= \frac{1}{mn} \left[\tau_{mnst} + d_{mnst} - \frac{1}{(m+1)} \sum_{p=0}^{m} \tau_{pnst} - \frac{1}{(n+1)} \sum_{q=0}^{n} \tau_{mqst} \right],$$

which implies that

$$\tau_{mnst} + d_{mnst} - T_m(n) - T_n(m) = mn[d_{mnst} - d_{m-1,n,s,t} - d_{m,n-1,s,t} + d_{m-1,n-1,s,t}].$$

$\square$

Lemma 2.10 *Let* $V_{mnst} = \sum_{p=m}^{\infty} \sum_{q=n}^{\infty} |d_{pqst} - d_{p-1,q,s,t} - d_{p,q-1,s,t} + d_{p-1,q-1,s,t}|$. *Then*

$$V_{mnst} - V_{m,n+1,s,t} - V_{m+1,n,s,t} + V_{m+1,n+1,s,t}$$
$$= |d_{mnst} - d_{m-1,n,s,t} - d_{m,n-1,s,t} + d_{m-1,n-1,s,t}|.$$

Proof We have

$$V_{mnst} - V_{m,n+1,s,t} - V_{m+1,n,s,t} + V_{m+1,n+1,s,t}$$

$$= \sum_{p=m}^{\infty} \sum_{q=n}^{\infty} |d_{pqst} - d_{p-1,q,s,t} - d_{p,q-1,s,t} + d_{p-1,q-1,s,t}|$$

$$- \sum_{p=m}^{\infty} \sum_{q=n+1}^{\infty} |d_{pqst} - d_{p-1,q,s,t} - d_{p,q-1,s,t} + d_{p-1,q-1,s,t}|$$

$$- \sum_{p=m+1}^{\infty} \sum_{q=n}^{\infty} |d_{pqst} - d_{p-1,q,s,t} - d_{p,q-1,s,t} + d_{p-1,q-1,s,t}|$$

$$+ \sum_{p=m+1}^{\infty} \sum_{q=n+1}^{\infty} |d_{pqst} - d_{p-1,q,s,t} - d_{p,q-1,s,t} + d_{p-1,q-1,s,t}|$$

$$= \sum_{p=m}^{\infty} \left[\sum_{q=n}^{\infty} - \sum_{q=n+1}^{\infty} \right] |d_{pqst} - d_{p-1,q,s,t} - d_{p,q-1,s,t} + d_{p-1,q-1,s,t}|$$

$$- \sum_{p=m+1}^{\infty} \left[\sum_{q=n}^{\infty} - \sum_{q=n+1}^{\infty} \right] |d_{pqst} - d_{p-1,q,s,t} - d_{p,q-1,s,t} + d_{p-1,q-1,s,t}|$$

$$= \sum_{p=m}^{\infty} |d_{pnst} - d_{p-1,n,s,t} - d_{p,n-1,s,t} + d_{p-1,n-1,s,t}|$$

$$- \sum_{p=m+1}^{\infty} |d_{pnst} - d_{p-1,n,s,t} - d_{p,n-1,s,t} + d_{p-1,n-1,s,t}|$$

$$= \left[\sum_{p=m}^{\infty} - \sum_{p=m+1}^{\infty} \right] |d_{pnst} - d_{p-1,n,s,t} - d_{p,n-1,s,t} + d_{p-1,n-1,s,t}|$$

$$= |d_{mnst} - d_{m-1,n,s,t} - d_{m,n-1,s,t} + d_{m-1,n-1,s,t}|. \qquad \square$$

2.6 Inclusion Relations

We establish here some inclusion relations between the spaces defined in Sect. 2.3.

Theorem 2.11 *We have the following inclusions with the limit preserved in each case, while the reverse inclusions do not hold in general.*

(i) $\hat{w}_2 \subset [w_2]$ *if conditions* (ii) *and* (iii) *of Lemma* 2.8 *hold.*
(ii) $\hat{w}_2 \subset \hat{w}_2$.

Proof (i) We have to show that $\hat{w}_2 \subset [w_2]$. If $x \in \hat{w}_2$, then we have

$$V_{mnst} = \sum_{p=m}^{\infty} \sum_{q=n}^{\infty} |d_{pqst} - d_{p-1,q,s,t} - d_{p,q-1,s,t} + d_{p-1,q-1,s,t}| \qquad (2.25)$$

$\longrightarrow 0$ as $m, n \longrightarrow \infty$, uniformly in s, t, and

$$d_{pqst} \longrightarrow L \quad (\text{say}) \text{ as } p, q \longrightarrow \infty \text{ uniformly in } s, t,$$

that is, $w_2\text{-}\lim x = L$.

In order to prove that $x \in [w_2]$, it is enough to show that condition (iv) of Lemma 2.8 holds. By Lemmas 2.9 and 2.10 we have

$$\tau_{pqst} + d_{pqst} - T_1(p, q, s, t) - T_2(p, q, s, t)$$
$$= pq[d_{pqst} - d_{p-1,q,s,t} - d_{p,q-1,s,t} + d_{p-1,q-1,s,t}]$$

and

$$|d_{pqst} - d_{p-1,q,s,t} - d_{p,q-1,s,t} + d_{p-1,q-1,s,t}|$$
$$= V_{pqst} - V_{p,q+1,s,t} - V_{p+1,q,s,t} + V_{p+1,q+1,s,t},$$

so that

$$\frac{1}{mn} \sum_{p=1}^{m} \sum_{q=1}^{n} |\tau_{pqst} + d_{pqst} - T_1(p, q, s, t) - T_2(p, q, s, t)|$$

$$= \frac{1}{mn} \sum_{p=1}^{m} \sum_{q=1}^{n} pq[V_{pqst} - V_{p,q+1,s,t} - V_{p+1,q,s,t} + V_{p+1,q+1,s,t}]$$

using Lemma 2.7 for Abel's transformation, we have

$$= \frac{1}{mn} \left[\sum_{p=1}^{m} \sum_{q=1}^{n} V_{pqst} - m \sum_{q=1}^{n} V_{m+1,q,s,t} - n \sum_{p=1}^{m} V_{p,n+1,s,t} + mn V_{m+1,n+1,s,t} \right]$$

$\longrightarrow 0$ as $m, n \longrightarrow \infty$, uniformly in s, t (by 2.25). Hence, by Lemma 2.8, $x \in [w_2]$.

(ii) Let $x \in \hat{w}_2$. We have to show that

$$\sup_{s,t} \sum_{p=0}^{\infty} \sum_{q=0}^{\infty} |d_{pqst} - d_{p-1,q,s,t} - d_{p,q-1,s,t} + d_{p-1,q-1,s,t}| \leq K,$$

where K is an absolute constant. As $x \in \hat{w}_2$, there exist integers p_0, q_0 such that

$$\sum_{p>p_0} \sum_{q>q_0} |d_{pqst} - d_{p-1,q,s,t} - d_{p,q-1,s,t} + d_{p-1,q-1,s,t}| < 1, \quad \text{for all } s, t. \quad (2.26)$$

Hence, it is left to show that, for fixed p, q,

$$|d_{pqst} - d_{p-1,q,s,t} - d_{p,q-1,s,t} + d_{p-1,q-1,s,t}| \leq K \quad \text{for all } s, t.$$

From (2.26) we have that

$$|d_{pqst} - d_{p-1,q,s,t} - d_{p,q-1,s,t} + d_{p-1,q-1,s,t}| < 1 \quad (2.27)$$

for every fixed $p > p_0, q > q_0$ and all s, t. Since

$$m(m+1)n(n+1)(d_{mnst} - d_{m-1,n,s,t} - d_{m,n-1,s,t} + d_{m-1,n-1,s,t})$$

$$= \sum_{p=1}^{m} \sum_{q=1}^{n} pq(\tau_{pqst} - \tau_{p-1,q,s,t} - \tau_{p,q-1,s,t} + \tau_{p-1,q-1,s,t}), \quad (2.28)$$

we have

$$mn\big[(m+1)(n+1)(d_{mnst} - d_{m-1,n,s,t} - d_{m,n-1,s,t} + d_{m-1,n-1,s,t})$$

$$\quad - (m-1)(n+1)(d_{m-1,n,s,t} - d_{m-2,n,s,t} - d_{m-1,n-1,s,t} + d_{m-2,n-1,s,t})$$

$$\quad - (m+1)(n-1)(d_{m,n-1,s,t} - d_{m-1,n-1,s,t} - d_{m,n-2,s,t} + d_{m-1,n-2,s,t})$$

$$\quad + (m-1)(n-1)(d_{m-1,n-1,s,t} - d_{m-2,n-1,s,t} - d_{m-1,n-2,s,t} + d_{m-2,n-2,s,t}) \big]$$

$$= \sum_{q=1}^{n} q \left[\sum_{p=1}^{m} p(\tau_{pqst} - \tau_{p-1,q,s,t} - \tau_{p,q-1,s,t} + \tau_{p-1,q-1,s,t}) \right.$$

$$\left. - \sum_{p=1}^{m-1} p(\tau_{pqst} - \tau_{p-1,q,s,t} - \tau_{p,q-1,s,t} + \tau_{p-1,q-1,s,t}) \right]$$

$$-\sum_{q=1}^{n-1} q\left[\sum_{p=1}^{m} p(\tau_{pqst} - \tau_{p-1,q,s,t} - \tau_{p,q-1,s,t} + \tau_{p-1,q-1,s,t})\right.$$

$$\left.-\sum_{p=1}^{m-1} p(\tau_{pqst} - \tau_{p-1,q,s,t} - \tau_{p,q-1,s,t} + \tau_{p-1,q-1,s,t})\right]$$

$$=\sum_{q=1}^{n} q\left[m(\tau_{mqst} - \tau_{m-1,q,s,t} - \tau_{m,q-1,s,t} + \tau_{m-1,q-1,s,t})\right]$$

$$-\sum_{q=1}^{n-1} q\left[m(\tau_{mqst} - \tau_{m-1,q,s,t} - \tau_{m,q-1,s,t} + \tau_{m-1,q-1,s,t})\right].$$

This implies that

$$(m+1)(n+1)(d_{mnst} - d_{m-1,n,s,t} - d_{m,n-1,s,t} + d_{m-1,n-1,s,t})$$
$$- (m-1)(n+1)(d_{m-1,n,s,t} - d_{m-2,n,s,t} - d_{m-1,n-1,s,t} + d_{m-2,n-1,s,t})$$
$$- (m+1)(n-1)(d_{m,n-1,s,t} - d_{m-1,n-1,s,t} - d_{m,n-2,s,t} + d_{m-1,n-2,s,t})$$
$$+ (m-1)(n-1)(d_{m-1,n-1,s,t} - d_{m-2,n-1,s,t} - d_{m-1,n-2,s,t} + d_{m-2,n-2,s,t})$$
$$= [\tau_{mnst} - \tau_{m-1,n,s,t} - \tau_{m,n-1,s,t} + \tau_{m-1,n-1,s,t}]. \tag{2.29}$$

Using (2.27) and (2.29), we have that

$$|\tau_{mnst} - \tau_{m-1,n,s,t} - \tau_{m,n-1,s,t} + \tau_{m-1,n-1,s,t}| \le K(m,n) \tag{2.30}$$

for every fixed $m > p_0$, $n > q_0$ and all s, t, where $K(m,n)$ is a constant depending on m, n. Again, from the definition of τ_{mnst} we obtain, similarly to (2.25),

$$\tau_{mnst} - \tau_{m-1,n,s,t} - \tau_{m,n-1,s,t} + \tau_{m-1,n-1,s,t}$$

$$= \frac{1}{m(m+1)n(n+1)} \sum_{u=1}^{m} \sum_{v=1}^{n} uva_{u+s,v+t}, \tag{2.31}$$

so that

$$a_{m+s,n+t}$$
$$= (m+1)(n+1)(\tau_{mnst} - \tau_{m-1,n,s,t} - \tau_{m,n-1,s,t} + \tau_{m-1,n-1,s,t})$$
$$- (m-1)(n+1)(\tau_{m-1,n,s,t} - \tau_{m-2,n,s,t} - \tau_{m-1,n-1,s,t} + \tau_{m-2,n-1,s,t})$$
$$- (m+1)(n-1)(\tau_{m,n-1,s,t} - \tau_{m-1,n-1,s,t} - \tau_{m,n-2,s,t} + \tau_{m-1,n-2,s,t})$$
$$+ (m-1)(n-1)(\tau_{m-1,n-1,s,t} - \tau_{m-2,n-1,s,t} - \tau_{m-1,n-2,s,t} + \tau_{m-2,n-2,s,t}).$$

Hence, it follows from (2.30) that, for any fixed $m > p_0$ and $n > q_0$,

$$|a_{m+s,n+t}| \leq K(m,n) \quad \text{for all } s,t. \tag{2.32}$$

Now choose $m = m_0 + 1, n = n_0 + 1$. Let

$$K = \max\{K(m_0+1, n_0+1), |a_{1,n_0+1}|, |a_{m_0+1,1}|, |a_{2,n_0+1}|, |a_{m_0+1,2}|, \ldots,$$
$$|a_{n_0+1,n_0+1}|\}.$$

It follows from (2.32) that

$$|a_{u,v}| \leq K \quad \text{for all } u, v$$

where K is independent of u, v. By (2.31) we have

$$|\tau_{mnst} - \tau_{m-1,n,s,t} - \tau_{m,n-1,s,t} + \tau_{m-1,n-1,s,t}| \leq K \quad \text{for all } m,n,s,t. \tag{2.33}$$

Also, from (2.28) and (2.33) we have

$$|d_{pqst} - d_{p-1,q,s,t} - d_{p,q-1,s,t} + d_{p-1,q-1,s,t}| \leq K \quad \text{for all } p,q,s,t. \qquad \square$$

Theorem 2.12 *We have the following proper inclusions with the limit preserved in each case*:

$$[\mathcal{F}] \subset [w]_2 \subset [w_2] \subset w_2 \subset (C_2, 2).$$

Proof Let $x \in [\mathcal{F}]$ with $[\mathcal{F}]\text{-}\lim x = L$, say. Then

$$\tau_{pqst}(|x - Le|) \longrightarrow 0 \quad \text{as } p,q \longrightarrow \infty, \text{ uniformly in } s, t.$$

This implies that

$$\frac{1}{(m+1)(n+1)} \sum_{p=0}^{m} \sum_{q=0}^{n} \tau_{pqst}(|x - Le|) \longrightarrow 0 \quad \text{as } p,q \longrightarrow \infty, \text{ uniformly in } s,t.$$

This proves that $x \in [w]_2$ and $[\mathcal{F}]\text{-}\lim x = [w]_2\text{-}\lim x = L$.

Since

$$\frac{1}{(m+1)(n+1)} \left| \sum_{p=0}^{m} \sum_{q=0}^{n} \tau_{pqst}(x - Le) \right| \leq \frac{1}{(m+1)(n+1)} \sum_{p=0}^{m} \sum_{q=0}^{n} |\tau_{pqst}(x - Le)|$$

$$\leq \frac{1}{(m+1)(n+1)} \sum_{p=0}^{m} \sum_{q=0}^{n} \tau_{pqst}(|x - Le|),$$

this implies that $[w]_2 \subset [w_2] \subset w_2$ and

$$[w]_2\text{-}\lim x = [w_2]\text{-}\lim x = w_2\text{-}\lim x = L.$$

Since

$$\frac{1}{(m+1)(n+1)} \sum_{p=0}^{m} \sum_{q=0}^{n} \tau_{pqst}(x - \ell\mathbf{e})$$

converges uniformly in s, t as $m, n \longrightarrow \infty$, we have the convergence for $s = 0 = t$. It follows that $w_2 \subset (C_2, 2)$ and $w_2\text{-}\lim x = (C_2, 2)\text{-}\lim x = L$.   □

Example 2.13 $[w]_2 \cap \mathcal{M}_u \subsetneq [w_2] \cap \mathcal{M}_u$.
Let $x = (x_{jk})$ be defined by

$$x_{jk} = (-1)^k \quad \text{for all } j,$$

that is,

$$\begin{pmatrix} -1 & 1 & -1 & 1 & \cdot & \cdot & \cdot \\ -1 & 1 & -1 & 1 & \cdot & \cdot & \cdot \\ -1 & 1 & -1 & 1 & \cdot & \cdot & \cdot \\ \cdot & \cdot & \cdot & \cdot & \cdot & \cdot & \cdot \\ \cdot & \cdot & \cdot & \cdot & \cdot & \cdot & \cdot \end{pmatrix}.$$

Then

$$\left| \tau_{pqst}(x - 0) \right| = \left| \frac{1}{(p+1)(q+1)} \sum_{j=0}^{p} \sum_{k=0}^{q} x_{j+s,k+t} \right|$$

$$\leq \frac{q+1}{(p+1)(q+1)} = \frac{1}{p+1} \quad \text{uniformly for } s, t.$$

Hence,

$$\frac{1}{(m+1)(n+1)} \sum_{p=0}^{m} \sum_{q=0}^{n} \left| \tau_{pqst}(x - 0) \right| \leq \frac{1}{(m+1)(n+1)} \sum_{p=0}^{m} \sum_{q=0}^{n} \frac{1}{p+1}$$

$$= \frac{1}{(n+1)(m+1)} \sum_{p=0}^{m} \frac{n+1}{p+1}$$

$$= \frac{1}{(m+1)} \sum_{p=0}^{m} \frac{1}{p+1} \to 0 \quad \text{as } m \to \infty,$$

i.e., $x = (x_{jk}) \in [w_2] \cap \mathcal{M}_u$, and hence, by Theorem 2.5, $x \in \mathcal{F}$. But $x \notin [w]_2 \cap \mathcal{M}_u$, and hence $x \notin [\mathcal{F}]$.

2.7 Exercises

1 Check whether the double sequence $x = (x_{nk})$ defined by

$$x_{nk} = \begin{cases} 1 & \text{if } n \text{ and } k \text{ are squares,} \\ 0 & \text{otherwise} \end{cases}$$

is almost convergent or not. If yes, then what is its almost limit?

2 Show that the inclusion

$$[\mathcal{F}] \subset \mathcal{F}$$

is proper.

3 Prove that

$$[w]_2 \cap \mathcal{M}_u = \big\{ x = (x_{jk}) : \theta(x - L\mathbf{e}) = 0 \text{ for some } L \big\}$$
$$= \big\{ x = (x_{jk}) : F(x - L\mathbf{e}) = 0 \text{ for all } F \in \{\mathcal{M}_u, \theta\} \text{ and some } L \big\}.$$

4 Prove that

$$[\mathcal{F}] \subset [w_2] \subset w_2$$

and

$$[w_2]\text{-}\lim x = w_2\text{-}\lim x = L.$$

5 Show that the inclusion

$$[w]_2 \subset [w_2]$$

is proper.

6 Prove the following proper inclusions:

$$[\mathcal{F}] \subset [w]_2 \cap \mathcal{M}_u \subset [w_2] \cap \mathcal{M}_u \subset \mathcal{F}.$$

Chapter 3
Almost Regular Matrices

The Silverman–Toeplitz theorem is a well-known theorem that states necessary and sufficient conditions to transform a convergent sequence into a convergent sequence leaving the limit invariant. This idea was extended to RH-regular matrices by using the notion of P-convergence (see [50] and [111]). In this chapter, we use the notion of almost convergence to define and characterize almost conservative, almost regular, strongly regular, and almost strongly regular four-dimensional matrices.

3.1 Introduction

Let Ω denote the vector space of all double sequences with the vector space operations defined coordinate-wise. Vector subspaces of Ω are called *double sequence spaces*. In addition to the above-mentioned double sequence spaces, we consider the double sequence space

$$\mathcal{L}_u := \left\{ x \in \Omega \mid \|x\|_1 := \sum_{j,k} |x_{jk}| < \infty \right\}$$

of all absolutely summable double sequences.

All considered double sequence spaces are supposed to contain

$$\Phi := \operatorname{span}\{e^{jk} \mid j, k \in \mathbb{N}\},$$

where

$$e_{i\ell}^{jk} = \begin{cases} 1 & \text{if } (j,k) = (i,\ell), \\ 0 & \text{otherwise.} \end{cases}$$

We denote the pointwise sums $\sum_{j,k} e^{jk}$, $\sum_j e^{jk}$ ($k \in \mathbb{N}$), and $\sum_k e^{jk}$ ($j \in \mathbb{N}$) by $\mathbf{e}$, $\mathbf{e}^k$, and $\mathbf{e}_j$, respectively.

M. Mursaleen, S.A. Mohiuddine, *Convergence Methods for Double Sequences and Applications*, DOI 10.1007/978-81-322-1611-7_3, © Springer India 2014

Let E be the space of double sequences converging with respect to a convergence notion v, F be a double sequence space, and $A = (a_{mnjk})$ be a four-dimensional matrix of scalars. Define the set

$$F_A^{(v)} := \left\{ x \in \Omega \mid [Ax]_{mn} := v\text{-}\sum_{j,k} a_{mnjk} x_{jk} \text{ exists, and } Ax := \left([Ax]_{mn}\right)_{m,n} \in F \right\}.$$

Then we say that A maps the space E into the space F if $E \subset F_A^{(v)}$ and denote by (E, F) the set of all four-dimensional matrices A that map E into F.

We say that a four-dimensional matrix A is C_v-*conservative* if $C_v \subset C_{vA}^{(v)}$ and C_v-*regular* if, in addition,

$$v\text{-}\lim Ax := v\text{-}\lim_{m,n}[Ax]_{mn} = v\text{-}\lim_{m,n} x_{mn} \quad (x \in C_v).$$

Note that a convergent double sequence need not be almost convergent. However, every bounded convergent double sequence is almost convergent, and every almost convergent double sequence is bounded.

Let $A = (a_{mnjk})$ $(j, k = 0, 1, \dots)$ be a double infinite matrix of real numbers. For all $m, n = 0, 1, \dots$, the sums

$$y_{mn} = \sum_{j=0}^{\infty} \sum_{k=0}^{\infty} a_{mnjk} x_{jk}$$

are called the A-means of the sequence $x = (x_{jk})$. We say that a sequence x is A-summable to the limit t if the A-means exist for all $m, n = 0, 1, \dots$ in the sense of Pringsheim's convergence:

$$P\text{-}\lim_{p,q \to \infty} \sum_{j=0}^{p} \sum_{k=0}^{q} a_{mnjk} x_{jk} = y_{mn} \quad \text{and} \quad P\text{-}\lim_{m,n \to \infty} y_{mn} = t.$$

We say that a matrix A is *bounded-regular* (*or RH-regular*) if every bounded-convergent sequence x is A-summable to the same limit and the A-means are also bounded.

Four-dimensional matrices that map every almost convergent double sequence into a BP-convergent double sequence with the same limit were considered by Móricz and Rhoades [83], four-dimensional matrices that map every almost convergent double sequence into an almost convergent double sequence with the same limit were characterized by Mursaleen [88], and in [128], authors characterized the almost C_v-conservative matrices, i.e., the four-dimensional matrices $A = (a_{mnjk})$ that map the double sequence space C_v into the space $\mathcal{F}$, where $v \in \{BP, R, P\}$.

3.2 Almost C_v-Conservative Matrices

In this section, we define and characterize almost C_v-conservative matrices, i.e., the four-dimensional matrices $A = (a_{mnjk})$ that map the double sequence space C_v into the space $\mathcal{F}$, where $v \in \{BP, R, P\}$ (see [128]).

To derive them, we apply characterizations of four-dimensional matrices from the class (C_v, C_{bp}) for $v \in \{BP, R, P\}$.

The conditions for a four-dimensional matrix to map the spaces C_{BP}, C_R, C_P into the space C_{BP} are well known (see, for example, [50] and [111]).

Theorem 3.1

(a) *The matrix $A = (a_{mnjk})$ is in (C_R, C_{BP}) if and only if the following conditions hold:*

  (i) $\sup_{m,n} \sum_{j,k} |a_{mnjk}| < \infty$,
  (ii) *the limit $BP\text{-}\lim_{m,n} a_{mnjk} = a_{jk}$ exists $(j, k \in \mathbb{N})$,*
  (iii) *the limit $BP\text{-}\lim_{m,n} \sum_{j,k} a_{mnjk} = v$ exists,*
  (iv) *the limit $BP\text{-}\lim_{m,n} \sum_{j} a_{mnjk_0} = u^{k_0}$ exists $(k_0 \in \mathbb{N})$,*
  (v) *the limit $BP\text{-}\lim_{m,n} \sum_{k} a_{mnj_0 k} = v_{j_0}$ exists $(j_0 \in \mathbb{N})$.*

  In this case, $a = (a_{jk}) \in \mathcal{L}_u$, (u^k), $(v_j) \in l_1$, and

$$BP\text{-}\lim_{m,n}[Ax]_{m,n} = \sum_{j,k} a_{jk} x_{jk} + \sum_{j}\left(v_j - \sum_{k} a_{jk}\right)x_j + \sum_{k}\left(u^k - \sum_{j} a_{jk}\right)x^k$$

$$+ \left(v + \sum_{j,k} a_{jk} - \sum_{j} v_j - \sum_{k} u^k\right) R\text{-}\lim_{m,n} x_{mn} \quad (x \in C_R).$$

(b) *The matrix $A = (a_{mnjk})$ is in (C_R, C_{BP}) and $BP\text{-}\lim Ax = R\text{-}\lim_{m,n} x_{mn}$ $(x \in C_R)$ if and only if conditions (i)–(v) hold with $a_{jk} = u^k = v_j = 0$ $(j, k \in \mathbb{N})$ and $v = 1$.*

Theorem 3.2

(a) *The matrix $A = (a_{mnjk})$ is in (C_{BP}, C_{BP}) if and only if it satisfies conditions (i), (ii), and (iii) of Theorem 3.1 and*

  (vi) $BP\text{-}\lim_{m,n} \sum_{j} |a_{mnjk_0} - a_{jk_0}| = 0$ $(k_0 \in \mathbb{N})$,
  (vii) $BP\text{-}\lim_{m,n} \sum_{k} |a_{mnj_0 k} - a_{j_0 k}| = 0$ $(j_0 \in \mathbb{N})$.

  In this case, $a = (a_{jk}) \in \mathcal{L}_u$, and

$$BP\text{-}\lim_{m,n}[Ax]_{m,n} = \sum_{j,k} a_{jk} x_{jk} + \left(v - \sum_{j,k} a_{jk}\right) BP\text{-}\lim_{m,n} x_{mn} \quad (x \in C_{bp}).$$

(b) *The matrix $A = (a_{mnjk})$ is in (C_{BP}, C_{BP}) and $BP\text{-}\lim Ax = BP\text{-}\lim_{m,n} x_{mn}$ $(x \in C_{BP})$ if and only if conditions (i), (ii), (iii) of Theorem 3.1 and (vi) and (vii) hold with $a_{jk} = 0$ $(j, k \in \mathbb{N})$ and $v = 1$.*

Theorem 3.3

(a) *The matrix $A = (a_{mnjk})$ is in (C_P, C_{BP}) if and only if conditions (i)–(iii) of Theorem 3.1 hold and*

 (viii) *for every $j \in \mathbb{N}$, there exists $K \in \mathbb{N}$ such that $a_{mnjk} = 0$ for $k > K$*
  $(m, n \in \mathbb{N})$,
 (ix) *for every $k \in \mathbb{N}$, there exists $J \in \mathbb{N}$ such that $a_{mnjk} = 0$ for $j > J$*
  $(m, n \in \mathbb{N})$,

 In this case, $a = (a_{jk}) \in \mathcal{L}_u$, $(a_{jk_0})_j$, $(a_{j_0k})_k \in \varphi$ $(j_0, k_0 \in \mathbb{N})$, and

$$BP\text{-}\lim_{m,n}[Ax]_{m,n} = \sum_{j,k} a_{jk}x_{jk} + \sum_{j}\left(v - \sum_{j,k} a_{jk}\right) P\text{-}\lim_{m,n} x_{mn} \quad (x \in C_P).$$

(b) *The matrix $A = (a_{mnjk})$ is in (C_P, C_{BP}) and $BP\text{-}\lim Ax = P\text{-}\lim_{m,n} x_{mn}$ $(x \in C_P)$ if and only if conditions (ii) and (iii) of Theorem 3.1 and (viii)–(ix) hold with $a_{jk} = 0$ $(j, k \in \mathbb{N})$ and $v = 1$.*

A four-dimensional matrix $A = (a_{mnjk})$ is said to be *almost C_v-conservative* if it transforms every v-convergent double sequence $x = (x_{jk})$ into the almost convergent double sequence, where $v \in \{P, BP, R\}$, that is, $A \in (C_v, \mathcal{F})$.

A four-dimensional matrix $A = (a_{mnjk})$ is said to be *almost C_v-regular* if it is almost C_v-conservative and $\mathcal{F}\text{-}\lim Ax = v\text{-}\lim x$ for every $x \in C_v$.

Comparing this definition with the definition of an almost regular four-dimensional matrix by Mursaleen and Savaş [103] we see that the authors in fact considered almost C_{BP}-regular matrices. Almost conservative and almost regular matrices for single sequences were characterized by King [55].

In this section we characterize the four-dimensional matrices $A \in (C_v, \mathcal{F})$, where $v \in \{BP, R, P\}$ [128].

Theorem 3.4

(a) *A matrix $A = (a_{mnjk})$ is almost C_{BP}-conservative if and only if the following conditions hold:*

 (i) *$\sup_{m,n} \sum_{j,k} |a_{mnjk}| =: M < \infty$,*
 (ii) *the limit $BP\text{-}\lim_{p,q} \alpha(j, k, p, q, s, t) = a_{jk}$ exists $(j, k \in \mathbb{N})$ uniformly in $s, t \in \mathbb{N}$,*
 (iii) *the limit $BP\text{-}\lim_{p,q} \sum_{j,k} \alpha(j, k, p, q, s, t) = u$ exists uniformly in $s, t \in \mathbb{N}$,*
 (iv) *the limit $BP\text{-}\lim_{p,q} \sum_k |\alpha(j, k, p, q, s, t) - a_{jk}| = 0$ exists $(j \in \mathbb{N})$ uniformly in $s, t \in \mathbb{N}$,*
 (v) *the limit $BP\text{-}\lim_{p,q} \sum_j |\alpha(j, k, p, q, s, t) - a_{jk}| = 0$ exists $(k \in \mathbb{N})$ uniformly in $s, t \in \mathbb{N}$,*

 where

$$\alpha(j, k, p, q, s, t) = \frac{1}{pq} \sum_{m=s}^{s+p-1} \sum_{n=t}^{t+q-1} a_{mnjk}.$$

In this case, $a = (a_{jk}) \in \mathcal{L}_u$, and

$$\mathcal{F}\text{-}\lim Ax = \sum_{j,k} a_{jk}x_{jk} + \left(u - \sum_{j,k} a_{jk}\right) BP\text{-}\lim_{i,l} x_{il}, \tag{3.1}$$

that is,

$$BP\text{-}\lim_{p,q} \sum_{j,k} \alpha(j,k,p,q,s,t)x_{jk} = \sum_{j,k} a_{jk}x_{jk} + \left(u - \sum_{j,k} a_{jk}\right) BP\text{-}\lim_{i,l} x_{il}$$

uniformly in $s,t \in \mathbb{N}$.

(b) *$A = (a_{mnjk})$ is almost $\mathcal{C}_{BP}$-regular, i.e., $A \in (\mathcal{C}_{BP}, \mathcal{F})_{\mathrm{reg}}$ if and only if conditions (i)–(v) hold with $a_{jk} = 0$ $(j,k \in \mathbb{N})$ and $u = 1$.*

Proof (a) (*Necessity.*) Let $A \in (\mathcal{C}_{BP}, \mathcal{F})$. Condition (i) follows since $(\mathcal{C}_{BP}, \mathcal{F}) \subset (\mathcal{C}_{BP}, \mathcal{M}_u)$ (see [50, § 5(5)]). Since e^{jk} and e are in $\mathcal{C}_{BP}$, conditions (ii) and (iii) respectively follow.

It is obvious that if $A \in (\mathcal{C}_{BP}, \mathcal{F})$, then the matrix $B^{st} := (b^{st}_{pqjk})_{p,q,j,k} := (\alpha(j,k,p,q,s,t))_{p,q,j,k}$ is in $(\mathcal{C}_{BP}, \mathcal{C}_{BP})$ for every $s,t \in \mathbb{N}$. In particular, the double sequence $b^{st} = (b^{st}_{jk})$ with $b^{st}_{jk} := BP\text{-}\lim_{p,q} b^{st}_{pqjk} = a_{jk}$ is in $\mathcal{L}_u$, and

$$BP\text{-}\lim_{p,q} \sum_{k} \left|b^{st}_{pqjk} - b^{st}_{jk}\right| = BP\text{-}\lim_{p,q} \sum_{k} \left|\alpha(j,k,p,q,s,t) - a_{jk}\right| = 0$$

for every $s,t \in \mathbb{N}$.

To verify conditions (iv) and (v), we need to prove that these limits are uniform in $s,t \in \mathbb{N}$. Suppose on the contrary that, for given $j_0 \in \mathbb{N}$,

$$BP\text{-}\limsup_{p,q} \sum_{k} \left|\alpha(j_0,k,p,q,s,t) - a_{j_0k}\right| \neq 0.$$

Then there exists $\varepsilon > 0$ and index sequences (p_i), (q_i) such that

$$\sup_{s,t} \sum_{k} \left|\alpha(j_0,k,p_i,q_i,s,t) - a_{j_0k}\right| \geq \varepsilon \quad (i \in \mathbb{N}).$$

So for every $i \in \mathbb{N}$, we can choose $s_i, t_i \in \mathbb{N}$ such that

$$\sum_{k} \left|\alpha(j_0,k,p_i,q_i,s_i,t_i) - a_{j_0k}\right| \geq \varepsilon \quad (i \in \mathbb{N}).$$

Since

$$\sum_{k} \left|\alpha(j_0,k,p_i,q_i,s_i,t_i)\right| \leq \sup_{m,n} \sum_{j,k} |a_{mnjk}| < \infty,$$

$(a_{jk}) \in \mathcal{L}_u$, and by (ii) going to a subsequence of (p_i, q_i, s_i, t_i) if necessary we may find an index sequence (k_i) such that

$$\sum_{k=1}^{k_i} |\alpha(j_0, k, p_i, q_i, s_i, t_i) - a_{j_0 k}| \leq \frac{\varepsilon}{8} \quad \text{and}$$

$$\sum_{k=k_{i+1}+1}^{\infty} |\alpha(j_0, k, p_i, q_i, s_i, t_i)| + \sum_{k=k_{i+1}+1}^{\infty} |a_{j_0 k}| \leq \frac{\varepsilon}{8} \quad (i \in \mathbb{N}).$$

So

$$\sum_{k=k_i+1}^{k_{i+1}} |\alpha(j_0, k, p_i, q_i, s_i, t_i) - a_{j_0 k}| \geq \frac{3\varepsilon}{4} \quad (i \in \mathbb{N}).$$

We define the double sequence $x = (x_{jk})$ by

$$x_{jk} = \begin{cases} (-1)^i \, \text{sgn}(\alpha(j_0, k, p_i, q_i, s_i, t_i) - a_{j_0 k}) & \text{for } k_i < k \leq k_{i+1} \ (i \in \mathbb{N}), \ j = j_0, \\ 0 & \text{for } j \neq j_0. \end{cases}$$

Then $x \in \mathcal{C}_{BP0}$ with $\|x\|_\infty \leq 1$, but for i even, we have

$$\frac{1}{p_i q_i} \sum_{m=s_i}^{s_i+p_i-1} \sum_{n=t_i}^{t_i+q_i-1} (Ax)_{mn} - \sum_{j,k} a_{jk} x_{jk}$$

$$= \sum_k \alpha(j_0, k, p_i, q_i, s_i, t_i) x_{j_0 k} - \sum_k a_{j_0 k} x_{j_0 k}$$

$$\geq \sum_{k=k_i+1}^{k_{i+1}} \left(\alpha(j_0, k, p_i, q_i, s_i, t_i) - a_{j_0 k}\right) x_{j_0 k} - \sum_{k=1}^{k_i} |\alpha(j_0, k, p_i, q_i, s_i, t_i) - a_{j_0 k}|$$

$$- \sum_{k=k_{i+1}+1}^{\infty} |\alpha(j_0, k, p_i, q_i, s_i, t_i)| - \sum_{k=k_{i+1}+1}^{\infty} |a_{j_0 k}|$$

$$\geq \sum_{k=k_i+1}^{k_{i+1}} |\alpha(j_0, k, p_i, q_i, s_i, t_i) - a_{j_0 k}| - \frac{\varepsilon}{8} - \frac{\varepsilon}{8} \geq \frac{3\varepsilon}{4} - \frac{\varepsilon}{4} = \frac{\varepsilon}{2}.$$

Analogously, for i odd, we get

$$\frac{1}{p_i q_i} \sum_{m=s_i}^{s_i+p_i-1} \sum_{n=t_i}^{t_i+q_i-1} (Ax)_{mn} - \sum_{j,k} a_{jk} x_{jk} \leq -\frac{\varepsilon}{2}.$$

Hence, $\frac{1}{pq} \sum_{m=s}^{s+p-1} \sum_{n=t}^{t+q-1} (Ax)_{mn}$ does not converge as $p, q \to \infty$ uniformly in $s, t \in \mathbb{N}$, that is, $Ax \notin \mathcal{F}$, giving the contradiction. Hence, (iv) holds. In the same way, we get that (v) holds.

(*Sufficiency.*) Let conditions (i)–(v) hold. Then, for any s, t, the matrix $B^{st} := (\alpha(j, k, p, q, s, t))_{p,q,j,k}$ is in $(\mathcal{C}_{BP}, \mathcal{C}_{BP})$. In particular,

$$BP\text{-}\lim_{p,q} \sum_{j,k} \alpha(j, k, p, q, s, t) x_{jk}$$

$$= \sum_{j,k} a_{jk} x_{jk} + \left(u - \sum_{j,k} a_{jk} \right) BP\text{-}\lim_{i,l} x_{il} \quad (s, t \in \mathbb{N}).$$

To prove that the limit is uniform in $s, t \in \mathbb{N}$, we consider

$$\sum_{j,k} \left(\alpha(j, k, p, q, s, t) - a_{jk} \right) \left(x_{jk} - BP\text{-}\lim_{i,l} x_{il} \right).$$

Let $\varepsilon > 0$ and $N \in \mathbb{N}$ be such that

$$\left| x_{jk} - BP\text{-}\lim_{i,l} x_{il} \right| \leq \frac{\varepsilon}{8M} \quad \text{for } j, k \geq N.$$

By (ii), (iv), and (v) we can choose $S \in \mathbb{N}$ such that for $p, q \geq S$ and every $s, t \in \mathbb{N}$, we have

$$\sum_{j=1}^{N-1} \sum_{k=1}^{N-1} \left| \alpha(j, k, p, q, s, t) - a_{jk} \right| \leq \frac{\varepsilon}{8\|x\|_\infty},$$

$$\sum_{j=1}^{N-1} \sum_{k} \left| \alpha(j, k, p, q, s, t) - a_{jk} \right| \leq \frac{\varepsilon}{8\|x\|_\infty},$$

$$\sum_{k=1}^{N-1} \sum_{j=N}^{\infty} \left| \alpha(j, k, p, q, s, t) - a_{jk} \right| \leq \frac{\varepsilon}{8\|x\|_\infty}.$$

Then

$$\left| \sum_{j,k} \left(\alpha(j, k, p, q, s, t) - a_{jk} \right) \left(x_{jk} - BP\text{-}\lim_{i,l} x_{il} \right) \right|$$

$$\leq 2 \sum_{j=1}^{N-1} \sum_{k=1}^{N-1} \left| \alpha(j, k, p, q, s, t) - a_{jk} \right| \|x\|_\infty$$

$$+ 2 \sum_{j=1}^{N-1} \sum_{k} \left| \alpha(j, k, p, q, s, t) - a_{jk} \right| \|x\|_\infty$$

$$+ 2 \sum_{k=1}^{N-1} \sum_{j=N}^{\infty} \left| \alpha(j, k, p, q, s, t) - a_{jk} \right| \|x\|_\infty$$

$$
+ \sum_{j=N}^{\infty} \sum_{k=N}^{\infty} \left(|\alpha(j,k,p,q,s,t)| + |a_{jk}| \right) \left| x_{jk} - BP\text{-}\lim_{i,l} x_{il} \right|
$$

$$
\leq \frac{\varepsilon}{4} + \frac{\varepsilon}{4} + \frac{\varepsilon}{4} + 2M \frac{\varepsilon}{8M} = \varepsilon \quad (s,t \in \mathbb{N}).
$$

Hence,

$$
BP\text{-}\lim_{p,q} \sum_{j,k} \left(\alpha(j,k,p,q,s,t) - a_{jk} \right) \left(x_{jk} - BP\text{-}\lim_{i,l} x_{il} \right) = 0
$$

uniformly in s, t, that is,

$$
BP\text{-}\lim_{p,q} \sum_{j,k} \alpha(j,k,p,q,s,t) x_{jk} = \sum_{j,k} a_{jk} x_{jk} + \left(u - \sum_{j,k} a_{jk} \right) BP\text{-}\lim_{i,l} x_{il}.
$$

(b) The sufficiency follows from (3.1), and the necessity follows from the inclusion $\{e^{jk}, e \mid j, k \in \mathbb{N}\} \subset \mathcal{C}_{BP}$. $\qquad\square$

Theorem 3.5

(a) *A matrix* $A = (a_{mnjk})$ *is almost* $\mathcal{C}_R$*-conservative if and only if conditions* (i)–(iii) *of Theorem* 3.4 *hold and*

   (iv) *the limit* $BP\text{-}\lim_{p,q} \sum_j \alpha(j,k_0,p,q,s,t) = u^{k_0}$ *exists uniformly in* s, t
   $(k_0 \in \mathbb{N})$,

   (v) *the limit* $BP\text{-}\lim_{p,q} \sum_k \alpha(j_0,k,p,q,s,t) = v_{j_0}$ *exists uniformly in* s, t
   $(j_0 \in \mathbb{N})$.

   In this case, $a = (a_{jk}) \in \mathcal{L}_u$, (u^k), $(v_j) \in l_1$, *and*

$$
\mathcal{F}\text{-}\lim Ax = \sum_{j,k} a_{jk} x_{jk} + \sum_j \left(v_j - \sum_k a_{jk} \right) x_j + \sum_k \left(u^k - \sum_j a_{jk} \right) x^k
$$

$$
+ \left(u + \sum_{j,k} a_{jk} - \sum_j v_j - \sum_k u^k \right) R\text{-}\lim x. \tag{3.2}
$$

(b) $A = (a_{mnjk})$ *is almost* $\mathcal{C}_v$*-regular, i.e.,* $A \in (\mathcal{C}_v, \mathcal{F})_{\mathrm{reg}}$, *if and only if conditions* (i)–(iii) *of Theorem* 3.4 *and* (iv), (v) *hold with* $u_{jk} = u^k = v_j = 0$
$(j, k \in \mathbb{N})$ *and* $u = 1$.

Proof (a) (*Necessity.*) Condition (i) holds since $(\mathcal{C}_R, \mathcal{F}) \subset (\mathcal{C}_R, \mathcal{M}_u)$. Conditions (ii), (iii), (iv), and (v) follow since $e^{jk}, e, e^k, e_j \in \mathcal{C}_R$ $(j, k \in \mathbb{N})$.

   (*Sufficiency.*) Let conditions (i)–(v) hold and first suppose that $x = (x_{jk}) \in \mathcal{C}_R$ satisfies $x_j = x^k = 0$ $(j, k \in \mathbb{N})$. Then also $R\text{-}\lim x = 0$.

By Theorem 3.1, the matrix $B^{st} := (\alpha(j,k,p,q,s,t))_{p,q,j,k}$ is in $(\mathcal{C}_R, \mathcal{C}_{BP})$ for any $s, t \in \mathbb{N}$. In particular,

$$BP\text{-}\lim_{p,q} \sum_{j,k} \alpha(j,k,p,q,s,t) x_{jk} = \sum_{j,k} a_{jk} x_{jk} \quad (s,t \in \mathbb{N}).$$

To prove that the limit is uniform in $s, t \in \mathbb{N}$, we consider

$$\sum_{j,k} (\alpha(j,k,p,q,s,t) - a_{jk}) x_{jk}.$$

Let $\varepsilon > 0$ and $N \in \mathbb{N}$ be such that

$$|x_{jk}| \leq \frac{\varepsilon}{4M} \quad \text{for } j \geq N \text{ or } k \geq N \ (j,k \in \mathbb{N}).$$

By (ii) we can choose $S \in \mathbb{N}$ such that for $p, q \geq S$ and any $s, t \in \mathbb{N}$,

$$\sum_{j=1}^{N-1} \sum_{k=1}^{N-1} |\alpha(j,k,p,q,s,t) - a_{jk}| \leq \frac{\varepsilon}{2\|x\|_\infty}.$$

Then

$$\left| \sum_{j,k} (\alpha(j,k,p,q,s,t) - a_{jk}) x_{jk} \right|$$

$$\leq \sum_{j=1}^{N-1} \sum_{k=1}^{N-1} |\alpha(j,k,p,q,s,t) - a_{jk}| \|x\|_\infty$$

$$+ \sum_{(j,k) \in \mathbb{N}^2 \setminus [1,N-1]^2} (|\alpha(j,k,p,q,s,t)| + |a_{jk}|) |x_{jk}|$$

$$\leq \frac{\varepsilon}{2} + 2M \frac{\varepsilon}{4M} = \varepsilon \quad (s,t \in \mathbb{N}).$$

Hence,

$$BP\text{-}\lim_{p,q} \sum_{j,k} (\alpha(j,k,p,q,s,t) - a_{jk}) x_{jk} = 0$$

uniformly in s, t, that is,

$$BP\text{-}\lim_{p,q} \sum_{j,k} \alpha(j,k,p,q,s,t) x_{jk} = \sum_{j,k} a_{jk} x_{jk}.$$

Now let $x = (x_{jk})$ be any element of $\mathcal{C}_R$ with $\xi := R\text{-}\lim x$. Then for the double sequence $z := (z_{jk})$ with $z_{jk} := x_{jk} - x_j - x^k + \xi$, we have $\lim_k z_{jk} = 0 \ (j \in \mathbb{N})$

and $\lim_j z_{jk} = 0$ $(k \in \mathbb{N})$. Hence,

$$BP\text{-}\lim_{p,q} \sum_{j,k} \alpha(j,k,p,q,s,t)\left(x_{jk} - x_j - x^k + \xi\right)$$

$$= BP\text{-}\lim_{p,q} \sum_{j,k} \alpha(j,k,p,q,s,t)z_{jk}$$

$$= \sum_{j,k} a_{jk}z_{jk} = \sum_{j,k} a_{jk}\left(x_{jk} - x_j - x^k + \xi\right).$$

The existence of the limit

$$BP\text{-}\lim_{p,q} \sum_{j,k} \alpha(j,k,p,q,s,t)x_{jk}$$

$$= \sum_{j,k} a_{jk}z_{jk} + BP\text{-}\lim_{p,q} \sum_{j,k} \alpha(j,k,p,q,s,t)(x_j - \xi)$$

$$+ BP\text{-}\lim_{p,q} \sum_{j,k} \alpha(j,k,p,q,s,t)\left(x^k - \xi\right) + BP\text{-}\lim_{p,q} \sum_{j,k} \alpha(j,k,p,q,s,t)\xi$$

then would follow if the limits on the right side exist.

The third limit

$$BP\text{-}\lim_{p,q} \sum_{j,k} \alpha(j,k,p,q,s,t)\xi = \xi v$$

exists by (iii).

We will show that the first limit equals $\sum_j v_j(x_j - \xi)$. To this end, let $\varepsilon > 0$ and $N \in \mathbb{N}$ be such that

$$|x_j - \xi| \leq \frac{\varepsilon}{4M} \quad \text{for } j \geq N.$$

By (v) we can choose $S \in \mathbb{N}$ such that for $p, q \geq S$ and any $s, t \in \mathbb{N}$,

$$\sum_{j=1}^{N-1} \left| \sum_k \alpha(j,k,p,q,s,t) - v_j \right| \leq \frac{\varepsilon}{4\|x\|_\infty}.$$

Then

$$\left| \sum_j \left(\sum_k \alpha(j,k,p,q,s,t) - v_j \right)(x_j - \xi) \right|$$

$$\leq 2 \sum_{j=1}^{N-1} \left| \sum_k \alpha(j,k,p,q,s,t) - v_j \right| \|x\|_\infty$$

$$+ \sum_{j=N}^{\infty} \left(\sum_k |\alpha(j,k,p,q,s,t)| + |v_j| \right) |x_{jk}|$$

$$\leq \frac{\varepsilon}{2} + 2M \frac{\varepsilon}{4M} = \varepsilon \quad (s,t \in \mathbb{N}).$$

Hence

$$BP\text{-}\lim_{p,q} \sum_{j,k} \alpha(j,k,p,q,s,t)(x_j - \xi) = \sum_j v_j(x_j - \xi).$$

Analogously,

$$BP\text{-}\lim_{p,q} \sum_{j,k} \alpha(j,k,p,q,s,t)\left(x^k - \xi\right) = \sum_k u^k\left(x^k - \xi\right).$$

Hence, the limit

$$BP\text{-}\lim_{p,q} \sum_{j,k} \alpha(j,k,p,q,s,t)x_{jk}$$

exists, and formula (3.2) holds.

(b) The sufficiency follows from (3.2), and the necessity follows from the inclusion $\{e^{jk}, e, e^k, e_j \mid j,k \in \mathbb{N}\} \subset C_R$.   □

Theorem 3.6

(a) *A matrix $A = (a_{mnjk})$ is almost C_P-conservative if and only if conditions (i)–(iii) of Theorem 3.4 and (viii), (ix) of Theorem 3.3 hold.*

   In this case, $a = (a_{jk}) \in \mathcal{L}_u$, $(a_{jk_0})_j$, $(a_{j_0k})_k \in \varphi$ $(j_0, k_0 \in \mathbb{N})$, and

$$\mathcal{F}\text{-}\lim Ax = \sum_{j,k} a_{jk}x_{jk} + \left(u - \sum_{j,k} a_{jk} \right) P\text{-}\lim_{i,l} x_{il}. \tag{3.3}$$

(b) *$A = (a_{mnjk})$ is almost C_P-regular, i.e., $A \in (C_P, \mathcal{F})_{\text{reg}}$ if and only if conditions (i)–(iii) of Theorem 3.4 and (viii), (ix) of Theorem 3.3 hold with $a_{jk} = 0$ $(j,k \in \mathbb{N})$ and $u = 1$.*

Proof (a) *Necessity* of conditions (i)–(iii) follows in the same way as in Theorem 3.1. Conditions (viii), (ix) of Theorem 3.3 follow since $(C_P, \mathcal{F}) \subset (C_P, \mathcal{M}_u)$ (see [50, § 5(6)]).

   (*Sufficiency.*) First, note that condition (viii) of Theorem 3.3 implies that $\alpha(j_0, k, p, q, s, t) = 0$ for given $j_0 \in$, $k > K$ and any $p, q, s, t \in \mathbb{N}$. Hence, also $a_{j_0k} = 0$ for $k > K$. Now, in view of (ii), condition (iv) of Theorem 3.4 follows. Analogously, condition (v) of Theorem 3.1 and $(a_{jk_0})_j \in \varphi$ $(k_0 \in \mathbb{N})$ follow from condition (ix) of Theorem 3.3. So, in view of Theorem 3.4, A is almost C_{BP}-conservative.

Now let $x \in \mathcal{C}_P$. Then there exists $N \in \mathbb{N}$ such that

$$\sup_{k,l>N} |x_{kl}| < \infty.$$

We consider x as a decomposition $x = y + z$ where y is an element of $\mathcal{C}_{BP}$ defined by $y_{kl} := x_{kl}$ for $k, l > N$ and $y_{kl} := 0$ for $k \leq N$ or $l \leq N$ and $z := x - y$. So, $Ay \in \mathcal{F}$, and

$$\mathcal{F}\text{-}\lim Ay = \sum_{j,k>N} a_{jk} x_{jk} + \left(u - \sum_{j,k} a_{jk} \right) P\text{-}\lim_{i,l} x_{il}.$$

To prove that $Ax \in \mathcal{F}$, we need to verify that $Az \in \mathcal{F}$. To this end, let $K \in \mathbb{N}$ be such that $a_{mnjk} = 0$ for $k > K$, $j = 1, \ldots, N$, and any $m, n \in \mathbb{N}$. Let also $J \in \mathbb{N}$ be such that $a_{mnjk} = 0$ for $j > J$, $k = 1, \ldots, N$, and any $m, n \in \mathbb{N}$. Then

$$Az = \sum_{j=1}^{N} \sum_{k=1}^{K} z_{jk} Ae^{jk} + \sum_{k=1}^{N} \sum_{j=N+1}^{J} z_{jk} Ae^{jk} \in \mathcal{F}$$

and

$$\mathcal{F}\text{-}\lim Az = \sum_{j=1}^{N} \sum_{k=1}^{K} a_{jk} z_{jk} + \sum_{k=1}^{N} \sum_{j=N+1}^{J} a_{jk} z_{jk} = \sum_{j=1}^{N} \sum_{k} a_{jk} z_{jk} + \sum_{k=1}^{N} \sum_{j=N+1}^{\infty} a_{jk} z_{jk}.$$

Hence, $Ax = Ay + Az \in \mathcal{F}$, and formula (3.3) holds.

(b) can be proved in the same way as in Theorem 3.4.   □

3.3 Strongly Regular Matrices

The notion of strong regularity for single sequences was introduced by Lorentz [63], and for double sequences, by Móricz and Rhoades [83]. We say that a four-dimensional matrix A is *strongly regular* if it maps every almost convergent double sequence $x = (x_{jk})$ into a *BP*-convergent double sequence with $\mathcal{F}\text{-}\lim x = BP\text{-}\lim Ax$, i.e., $A \in (\mathcal{F}, \mathcal{C}_{BP})_{\text{reg}}$. Necessary and sufficient conditions were also established for a matrix $A = (a_{mnjk})$ to be strongly regular in [83] as follows.

Theorem 3.7 *A matrix* $A = (a_{mnjk})$ *is strongly regular if and only if* A *is in* $(\mathcal{C}_{BP}, \mathcal{C}_{BP})$ *with* $BP\text{-}\lim Ax = BP\text{-}\lim_{m,n} x_{mn}$ $(x \in \mathcal{C}_{BP})$ *and satisfies the following two conditions*:

$$\lim_{m,n\to\infty} \sum_{j=0}^{\infty} \sum_{k=0}^{\infty} |\Delta_{10} a_{mnjk}| = 0, \qquad (3.4)$$

$$\lim_{m,n\to\infty}\sum_{j=0}^{\infty}\sum_{k=0}^{\infty}|\triangle_{01}\,a_{mnjk}|=0, \tag{3.5}$$

where $\triangle_{10}a_{mnjk}=a_{mnjk}-a_{m,n,j+1,k}$ and $\triangle_{01}a_{mnjk}=a_{mnjk}-a_{m,n,j,k+1}$ $(j,k,m,n=0,1,\dots)$.

Proof (Sufficiency.) Let x be an almost convergent sequence with limit s. We need to show that (y_{mn}) is convergent and bounded. Fix $\epsilon>0$. Then, from the definition of almost convergence, there exist fixed integers $p,q\geq 2$ such that

$$\frac{1}{pq}\sum_{j=m}^{m+p-1}\sum_{k=n}^{n+p-1}x_{jk}=s+\eta_{mn},$$

where

$$|\eta_{mn}|\leq\epsilon\quad(m,n=0,1,\dots). \tag{3.6}$$

Hence,

$$\sum_{pq}^{MN}=\frac{1}{pq}\sum_{m=0}^{\infty}\sum_{n=0}^{\infty}a_{MNmn}\sum_{j=m}^{m+p-1}\sum_{k=n}^{n+q-1}x_{jk}=sA_{MN}+\sum_{m=0}^{\infty}\sum_{n=0}^{\infty}a_{MNmn}\eta_{mn}, \tag{3.7}$$

where

$$A_{MN}=\sum_{m=0}^{\infty}\sum_{n=0}^{\infty}a_{MNmn}.$$

We will rewrite the left-hand side of (3.7) by interchanging the summations with respect to m and n with those with respect to j and k:

$$\sum_{pq}^{MN}=\frac{1}{pq}\sum_{j=0}^{p-2}\sum_{k=0}^{q-2}x_{jk}\sum_{m=0}^{j}\sum_{n=0}^{k}a_{MNmn}+\frac{1}{pq}\sum_{j=p-1}^{\infty}\sum_{k=0}^{q-2}x_{jk}\sum_{m=j-p+1}^{j}\sum_{n=0}^{k}a_{MNmn}$$

$$+\frac{1}{pq}\sum_{j=0}^{p-2}\sum_{k=q-1}^{\infty}x_{jk}\sum_{m=0}^{j}\sum_{n=k-q+1}^{k}a_{MNmn}$$

$$+\frac{1}{pq}\sum_{j=p-1}^{\infty}\sum_{k=q-1}^{\infty}x_{jk}\sum_{m=j-p+1}^{j}\sum_{n=k-q+1}^{k}a_{MNmn}. \tag{3.8}$$

The following estimates can be easily carried out. Because of condition (ii) of Theorem 3.2(b),

$$\left|\frac{1}{pq}\sum_{j=0}^{p-2}\sum_{k=0}^{q-2}x_{jk}\sum_{m=0}^{j}\sum_{n=0}^{k}a_{MNmn}\right|$$

$$\leq \frac{(p-1)(q-1)\|x\|}{pq} \sum_{j=0}^{p-2} \sum_{k=0}^{q-2} |a_{MNmn}|$$

$$\leq \|x\| \sum_{j=0}^{p-2} \sum_{k=0}^{q-2} |a_{MNmn}| \leq \epsilon \quad \text{if } \min(M, N) \geq M_1, \text{ say.} \qquad (3.9)$$

Because of condition (vi) of Theorem 3.2(b),

$$\frac{1}{pq} \sum_{j=p-1}^{\infty} \sum_{k=0}^{q-2} x_{jk} \sum_{m=j-p+1}^{j} \sum_{n=0}^{k} a_{MNmn}$$

$$\leq \frac{(q-1)\|x\|}{pq} \sum_{j=p-1}^{\infty} \sum_{m=j-p+1}^{j} \sum_{n=0}^{q-2} |a_{MNmn}|$$

$$\leq \|x\| \sum_{m=0}^{\infty} \sum_{n=0}^{q-2} |a_{MNmn}| \leq \epsilon \quad \text{if } \min(M, N) \geq M_2, \text{ say.} \qquad (3.10)$$

Similarly, because of condition (vii) of Theorem 3.2(b),

$$\left| \frac{1}{pq} \sum_{j=0}^{p-2} \sum_{k=q-1}^{\infty} x_{jk} \sum_{m=0}^{j} \sum_{n=k-q+1}^{\infty} a_{MNmn} \right| \leq \epsilon \quad \text{if } \min(M, N) \geq M_3, \text{ say.} \quad (3.11)$$

Finally, we can write

$$\frac{1}{pq} \sum_{j=p-1}^{\infty} \sum_{k=q-1}^{\infty} x_{jk} \sum_{m=j-p+1}^{j} \sum_{n=k-q+1}^{k} a_{MNmn}$$

$$= \sum_{j=p-1}^{\infty} \sum_{k=q-1}^{\infty} x_{jk} \left\{ \frac{1}{pq} \sum_{m=j-p+1}^{j} \sum_{n=k-q+1}^{k} a_{MNmn} - a_{MNjk} \right\}$$

$$+ y_{MN} - \sum_{j=0}^{p-2} \sum_{k=0}^{q-2} a_{MNjk} x_{jk} - \sum_{j=p-1}^{\infty} \sum_{k=0}^{q-2} a_{MNjk} x_{jk} - \sum_{j=0}^{p-2} \sum_{k=q-1}^{\infty} a_{MNjk} x_{jk}.$$

$$(3.12)$$

Here, again because of (ii), (vi), and (vii) of Theorem 3.2(b),

$$\left| \sum_{j=0}^{p-2} \sum_{k=0}^{q-2} a_{MNjk} x_{jk} \right| \leq \|x\| \sum_{j=0}^{p-2} \sum_{k=0}^{q-2} |a_{MNjk}| \leq \epsilon \quad \text{if } \min(M, N) \geq M_1, \quad (3.13)$$

$$\left|\sum_{j=p-1}^{\infty}\sum_{k=0}^{q-2}a_{MNjk}x_{jk}\right| \le \|x\|\sum_{j=p-1}^{\infty}\sum_{k=0}^{q-2}|a_{MNjk}| \le \epsilon \quad \text{if } \min(M,N) \ge M_2,$$

$$(3.14)$$

$$\left|\sum_{j=0}^{p-2}\sum_{k=q-1}^{\infty}a_{MNjk}x_{jk}\right| \le \epsilon \quad \text{if } \min(M,N) \ge M_3. \tag{3.15}$$

Our last aim is to show that the first double sum on the right-hand side of (3.12) is also as small as we wish if $M, N \to \infty$. To accomplish this, we estimate as follows:

$$\left|\sum_{j=p-1}^{\infty}\sum_{k=q-1}^{\infty}x_{jk}\left\{\frac{1}{pq}\sum_{m=j-p+1}^{j}\sum_{n=k-q+1}^{k}a_{MNmn} - a_{MNjk}\right\}\right|$$

$$\le \frac{\|x\|}{pq}\sum_{j=p-1}^{\infty}\sum_{k=q-1}^{\infty}\left|\sum_{m=j-p+1}^{j}\sum_{n=k-q+1}^{k}a_{mn}^{MN} - pqa_{jk}^{MN}\right|$$

$$\le \frac{\|x\|}{pq}\sum_{j=p-1}^{\infty}\sum_{k=q-1}^{\infty}\sum_{m=j-p+1}^{j}\sum_{n=k-q+1}^{k}|a_{MNmn} - a_{MNjk}|$$

$$\le \frac{\|x\|}{pq}\sum_{j=p-1}^{\infty}\sum_{k=q-1}^{\infty}\sum_{\pi=0}^{p-1}\sum_{\rho=0}^{q-1}|a_{M,N,\pi+j-p+1,\rho+k-q+1} - a_{MNjk}|$$

$$\le \frac{\|x\|}{pq}\sum_{\pi=0}^{p-1}\sum_{\rho=0}^{q-1}\sum_{j=p-1}^{\infty}\sum_{k=q-1}^{\infty}|a_{M,N,\pi+j-p+1,\rho+k-q+1} - a_{MNjk}|$$

$$\le \frac{\|x\|}{pq}\sum_{\pi=0}^{p-1}\sum_{\rho=0}^{q-1}\left\{(p-\pi-1)\sum_{j=0}^{\infty}\sum_{k=0}^{\infty}|\Delta_{10}a_{MNjk}|\right.$$

$$\left. + (q-\rho-1)\sum_{j=0}^{\infty}\sum_{k=0}^{\infty}|\Delta_{01}a_{MNjk}|\right\}$$

$$\le \epsilon \quad \text{if } \min(M,N) \ge M_4, \text{ say,} \tag{3.16}$$

due to (3.4) and (3.5), since p and q are fixed. Here we have used the representation (omitting the superscripts M and N)

$$a_{\pi+j-p+1,\rho+k-q+1} - a_{ij} = \sum_{\sigma=\pi+j-p+1}^{j-1}\Delta_{10}a_{\sigma,\rho+k-q+1} + \sum_{\tau=\rho+k-q+1}^{k-1}\Delta_{01}a_{j\tau}.$$

Collecting (3.7) and (3.16) together leads to the estimate

$$\left| y_{MN} - s A_{MN} - \sum_{m=0}^{\infty} \sum_{n=0}^{\infty} a_{MNmn} \eta_{mn} \right| \leq 7\epsilon$$

$$\text{if } \min(M, N) \geq \max(M_1, M_2, M_3, M_4). \tag{3.17}$$

Now we take into account conditions (i) and (iii) of Theorem 3.2(b) and (3.6) to obtain

$$|s A_{MN} - s| = |s| \left| \sum_{m=0}^{\infty} \sum_{n=0}^{\infty} a_{MNmn} - 1 \right| \leq |s| \epsilon \quad \text{if } \min(M, N) \geq \max M_5, \text{ say,}$$

and

$$\left| \sum_{m=0}^{\infty} \sum_{n=0}^{\infty} a_{MNmn} \eta_{mn} \right| \leq \epsilon \sum_{m=0}^{\infty} \sum_{n=0}^{\infty} |a_{MNmn}| \leq \epsilon C \quad (M, N = 0, 1, \dots).$$

Combining (3.17) with the last two inequalities provides

$$\left| y_{MN} - s \right| \leq \left(7 + |s| + C \right) \epsilon \quad \text{if } \min(M, N) \geq \max(M_1, M_2, M_3, M_4, M_5).$$

Since $\epsilon > 0$ is arbitrary, this proves the convergence of the A-means y_{mn} to s as M and N tend to ∞, that is, the A-summability of x to the same limit. The boundedness of (y_{MN}) follows from (i) of Theorem 3.2(b) and the boundedness of x.

(*Necessity.*) Let $A = (a_{mnjk})$ be a strongly regular matrix. Then A satisfies conditions of Theorem 3.2(b). First, we deal with the case where property (3.4) is not satisfied. Then there exists $\epsilon > 0$ such that, for infinitely many pairs of M and N both tending to ∞, we have

$$\sum_{m=0}^{\infty} \sum_{n=0}^{\infty} |\Delta_{10} a_{MNmn}| \geq 12\epsilon.$$

For any such M and N, we have either

$$\sum_{j=0}^{\infty} \sum_{k=0}^{\infty} |\Delta_{10} a_{M,N,2j,k}| \geq 6\epsilon \tag{3.18}$$

or

$$\sum_{j=0}^{\infty} \sum_{k=0}^{\infty} |\Delta_{10} a_{M,N,2j+1,k}| \geq 6\epsilon.$$

Without loss of generality, we may assume that the first of these inequalities, i.e., (3.18), is satisfied.

By recurrence we now construct three strictly increasing sequences (M_r), (N_r), and (p_r) of nonnegative integers, making use of condition (i) of Theorem 3.2(b) and (3.18). First, we set $p_0 = 0$ and choose M_1, N_1, and p_1 such that

$$\sum_{j=0}^{p_1-1} \sum_{k=0}^{p_1-1} |\Delta_{10}\, a_{M_1,N_1,2j,k}| \geq 5\epsilon \quad \text{and} \quad \sum_{j=2p_1}^{\infty} \sum_{k=p_1}^{\infty} |\Delta_{10}\, a_{M_1,N_1,j,k}| \geq \epsilon.$$

If the numbers M_ρ, N_ρ, and p_ρ are already defined for $\rho = 1, 2, \ldots, r-1$, then we choose M_r, N_r, and p_r such that

$$\sum_{j=0}^{p_{r-1}-1} \sum_{k=0}^{p_{r-1}-1} |a_{M_r,N_r,j,k}| \leq \epsilon, \quad \sum_{j=p_{r-1}}^{p_r-1} \sum_{k=p_{r-1}}^{p_r-1} |\Delta_{10}\, a_{M_r,N_r,j,k}| \geq 5\epsilon, \quad \text{and}$$

$$\sum_{j=2p_r}^{\infty} \sum_{k=p_r}^{\infty} |a_{M_r,N_r,j,k}| \leq \epsilon.$$

These choices are possible due to (i) and (ii) of Theorem 3.2(b) and (3.18).

Now we define the sequence (x_{jk}) as follows. Let

$$x_{2p_{r-1}+2j,\, p_{r-1}+k} = \left(-1^r\right) \operatorname{sgn} \Delta_{10} a_{M_r,N_r,2p_{r-1}+2j,\, p_{r-1}+k};$$

$$x_{2p_{r-1}+2j+1,\, p_{r-1}+k} = -x_{2p_{r-1}+2j,\, p_{r-1}+k}$$

for $j, k = 0, 1, \ldots, p_r - p_{r-1} - 1$ and $r = 1, 2, \ldots$; otherwise, let $x_{jk} = 0$. It is not hard to see that

$$|y_{M_r,N_r}| = \left| \sum_{j=0}^{\infty} \sum_{k=0}^{\infty} a_{M_r,N_r,j,k} x_{jk} \right|$$

$$\geq \sum_{j=p_{r-1}}^{p_r-1} \sum_{k=p_{r-1}}^{p_r-1} |\Delta_{10}\, a_{M_r,N_r,2j,k}| - 2 \sum_{j=0}^{2p_{r-1}-1} \sum_{k=0}^{p_{r-1}-1} |a_{M_r,N_r,j,k}|$$

$$- 2 \sum_{j=2p_r}^{\infty} \sum_{k=p_r}^{\infty} |a_{M_r,N_r,j,k}| \geq 5\epsilon - 2\epsilon - 2\epsilon = \epsilon$$

for $r = 2, 3, \ldots$ and that $\operatorname{sgn} y_{M_r,N_r} = (-1)^r$ for $r = 1, 2, \ldots$. We note that for $r = 1$, a similar argument yields $|y_{M_1,N_1}| \geq 3\epsilon$.

Hence, it follows that the sequence (y_{MN}) diverges as $M, N \to \infty$. On the other hand, it is easy to check that the sequence (x_{ik}) is bounded and almost convergent to 0. This proves the necessity of condition (3.4).

The necessity of (3.5) is established in a similar manner.   □

3.4 Almost Strongly Regular Matrices

In [88], Mursaleen introduced the concept of almost strongly regular matrices and found a set of necessary and sufficient conditions for $A = (a_{mnjk})$ to be almost strongly regular.

A four-dimensional matrix $A = (a_{mnjk})$ is said to be *almost strongly regular* (see [88]) if for every $x = (x_{jk}) \in \mathcal{F}$, $Ax \in \mathcal{F}$ with $\mathcal{F}\text{-}\lim Ax = \mathcal{F}\text{-}\lim x$, i.e., $A \in (\mathcal{F}, \mathcal{F})_{reg}$. Note that, in this case, A-means are obviously bounded since $Ax \in \mathcal{F}$, and every almost convergent double sequence is also bounded. Almost strongly regular matrices for single sequences were introduced and characterized by Duran [39].

First, we prove the following useful lemma.

Lemma 3.8 *Let $A = (a_{mnjk})$ be an almost regular matrix. Then there exists $x = (x_{jk}) \in \mathcal{M}_u$ such that $\|x\| \leq 1$ and*

$$\limsup_{p,q \to \infty} \sup_{s,t \geq 0} \sum_{j=0}^{\infty} \sum_{k=0}^{\infty} \alpha(j,k,p,q,s,t)x_{jk} = \limsup_{p,q \to \infty} \sup_{s,t \geq 0} \sum_{j=0}^{\infty} \sum_{k=0}^{\infty} |\alpha(j,k,p,q,s,t)|.$$

Proof Let

$$\lambda = \limsup_{p,q \to \infty} \sup_{s,t \geq 0} \sum_{j=0}^{\infty} \sum_{k=0}^{\infty} |\alpha(j,k,p,q,s,t)|,$$

and let for a given $\epsilon > 0$,

$$N(\epsilon) = \left\{ p,q \in \mathbb{N} : \sup_{s,t \geq 0} \sum_{j=0}^{\infty} \sum_{k=0}^{\infty} |\alpha(j,k,p,q,s,t)| > \lambda - \epsilon \right\}.$$

Then there exist increasing sequences of integers $p_r, q_r \in N(1/r)$ and j_r, k_r such that

$$\begin{cases} \sup\limits_{s,t \geq 0} \sum\limits_{j \leq j_{r-1}} \sum\limits_{k=0}^{\infty} |\alpha(j,k,p_r,q_r,s,t)| < 1/r, \\[2mm] \sup\limits_{s,t \geq 0} \sum\limits_{j > j_r} \sum\limits_{k=0}^{\infty} |\alpha(j,k,p_r,q_r,s,t)| < 1/r, \\[2mm] \sup\limits_{s,t \geq 0} \sum\limits_{j=0}^{\infty} \sum\limits_{k \leq k_{r-1}} |\alpha(j,k,p_r,q_r,s,t)| < 1/r, \\[2mm] \sup\limits_{s,t \geq 0} \sum\limits_{j=0}^{\infty} \sum\limits_{k > k_r} |\alpha(j,k,p_r,q_r,s,t)| < 1/r. \end{cases} \qquad (3.19)$$

Now define $x \in \mathcal{M}_u$ such that $j_{r-1} < j < j_r$, $k_{r-1} < k < k_r$,

$$
x_{jk} = \begin{cases} 1 & \text{if } \alpha(j,k,p_r,q_r,s,t) \geq 0, \\ -1 & \text{if } \alpha(j,k,p_r,q_r,s,t) < 0, \end{cases}
$$

$s,t = 1, 2, \ldots$. Then, for all $p_r, q_r \in N(1/r)$,

$$
\sum_j \sum_k \alpha(j,k,p_r,q_r,s,t) x_{jk}
$$

$$
= \left(\sum_{j \leq j_{r-1}} \sum_k + \sum_{j_{r-1} < j \leq j_r} \sum_k + \sum_{j > j_r} \sum_k + \sum_j \sum_{k \leq k_{r-1}} + \sum_j \sum_{k_{r-1} < k \leq k_r} \right.
$$

$$
\left. + \sum_j \sum_{k > k_r} \right) \alpha(j,k,p_r,q_r,s,t) x_{jk},
$$

$$
\geq \sum_{j_{r-1} < j \leq j_r} \sum_k \alpha(j,k,p_r,q_r,s,t) x_{jk} - \|x\| \sum_{j \leq j_{r-1}} \sum_k |\alpha(j,k,p_r,q_r,s,t)|
$$

$$
- \|x\| \sum_{j > j_r} \sum_k |\alpha(j,k,p_r,q_r,s,t)| + \sum_j \sum_{k_{r-1} < k \leq k_r} \alpha(j,k,p_r,q_r,s,t) x_{jk}
$$

$$
- \|x\| \sum_j \sum_{k \leq k_{r-1}} |\alpha(j,k,p_r,q_r,s,t)| - \|x\| \sum_j \sum_{k > k_r} |\alpha(j,k,p_r,q_r,s,t)|.
$$

Hence,

$$
\sup_{s,t} \sum_j \sum_k \alpha(j,k,p_r,q_r,s,t) x_{jk}
$$

$$
\geq \sup_{s,t} \sum_{j_{r-1} < j \leq j_r} \sum_k \alpha(j,k,p_r,q_r,s,t) x_{jk}
$$

$$
+ \sup_{s,t} \sum_j \sum_{k_{r-1} < k \leq k_r} \alpha(j,k,p_r,q_r,s,t) x_{jk} - 4/r,
$$

$$
= \sup_{s,t} \sum_{j_{r-1} < j \leq j_r} \sum_k |\alpha(j,k,p_r,q_r,s,t)|
$$

$$
+ \sup_{s,t} \sum_j \sum_{k_{r-1} < k \leq k_r} |\alpha(j,k,p_r,q_r,s,t)| - 4/r,
$$

$$
= \sup_{s,t} \left(\sum_j \sum_k - \sum_{j \leq j_{r-1}} \sum_k - \sum_{j > j_r} \sum_k - \sum_j \sum_{k \leq k_{r-1}} \right.
$$

$$
\left. - \sum_j \sum_{k > k_r} \right) |\alpha(j,k,p_r,q_r,s,t)| - 4/r
$$

$$\geq \sup_{s,t} \sum_j \sum_k |\alpha(j,k,p_r,q_r,s,t)| - 8/r.$$

Therefore,

$$\limsup_r \sup_{s,t} \sum_j \sum_k \alpha(j,k,p_r,q_r,s,t)x_{jk} \geq \lambda.$$

Also, for such x, it is obvious that

$$\limsup_r \sup_{s,t} \sum_j \sum_k \alpha(j,k,p_r,q_r,s,t)x_{jk} \leq \lambda,$$

whence the result.   □

In the following theorem, we characterize the almost strongly regular matrices, i.e., $A \in (\mathcal{F}, \mathcal{F})_{\text{reg}}$.

Theorem 3.9 *A matrix $A = (a_{mnjk})$ is almost strongly regular if and only if A is almost regular and satisfies the following two conditions*:

$$\lim_{p,q\to\infty} \sum_j \sum_k |\Delta_{10}\alpha(j,k,p,q,s,t)| = 0, \quad \text{uniformly in } s,t \geq 0, \qquad (3.20)$$

$$\lim_{p,q\to\infty} \sum_j \sum_k |\Delta_{01}\alpha(j,k,p,q,s,t)| = 0, \quad \text{uniformly in } s,t \geq 0, \qquad (3.21)$$

where

$$\Delta_{10}\alpha(j,k,p,q,s,t) = \alpha(j,k,p,q,s,t) - \alpha(j+1,k,p,q,s,t)$$

and

$$\Delta_{01}\alpha(j,k,p,q,s,t) = \alpha(j,k,p,q,s,t) - \alpha(j,k+1,p,q,s,t).$$

Proof (Sufficiency.) Let $x = (x_{jk})$ be an almost convergent sequence with $\mathcal{F}$-limit L. Then on the same lines as by Móricz and Rhoades [83], using the conditions of almost regularity and conditions (3.20) and (3.21), we can show that $y_{mn} \to L$ as $M, N \to \infty$ and that (y_{MN}) is bounded. Since a bounded convergent double sequence is also almost convergent to the same limit, we have $Ax \in \mathcal{F}$ and $\mathcal{F}$-lim $Ax = L$. Hence, A is almost strongly regular.

(*Necessity.*) Let $A \in (\mathcal{F}, \mathcal{F})_{\text{reg}}$, Since $(\mathcal{F}, \mathcal{F})_{\text{reg}} \subset (\mathcal{C}_{BP}, \mathcal{F})_{\text{reg}}$, we have $A \in (\mathcal{C}_{BP}, \mathcal{F})_{\text{reg}}$, i.e., A is almost regular. To prove the necessity of (3.20), let us define $x = (x_{jk})$ by

$$x_{2v_{r-1}+2j,v_{r-1}+k} = (-1)^r \operatorname{sgn} \Delta_{10}\alpha(2v_{r-1}+2j, v_{r-1}+k, p_r, q_r, s, t), \quad s,t > 0,$$

$$x_{2v_{r-1}+2j+1,v_{r-1}+k} = -x_{2v_{r-1}+2j,v_{r-1}+k}$$

for $j,k = 1,2,\ldots,v_r - v_{r-1}$ and $r = 1,2,\ldots$; otherwise, $x_{jk} = 0$.

Then x is bounded and almost convergent to 0; also $\|x\| \leq 1$. Since A is almost strongly regular, we have

$$\mathcal{F}\text{-}\lim Ax = \mathcal{F}\text{-}\lim x = 0.$$

Hence, by the above Lemma 3.8 we have

$$0 = \limsup_{p,q \to \infty} \sup_{s,t \geq 0} \sum_{j=0}^{\infty} \sum_{k=0}^{\infty} \Delta_{10}\alpha(j,k,p,q,s,t)x_{jk}$$

$$= \limsup_{p,q \to \infty} \sup_{s,t \geq 0} \sum_{j=0}^{\infty} \sum_{k=0}^{\infty} |\Delta_{10}\alpha(j,k,p,q,s,t)|,$$

i.e., (3.20) holds. Similarly, we can prove the necessity of (3.21). $\qquad \square$

3.5 Exercises

1 Prove Theorems 3.1, 3.2, and 3.3.

2 State and prove four-dimensional almost coercive matrices, i.e., $(\mathcal{M}_u, \mathcal{F})$.

3 Characterize the class $(\mathcal{F}, \mathcal{M}_u)$.

4 Characterize the class $(\mathcal{L}_u, \mathcal{F})$.

5 Analogously to Theorem 1.46, prove that $A = (a_{mnjk}) \in ([\mathcal{F}], \mathcal{C}_{BP})$ with $BP\text{-}\lim Ax = [\mathcal{F}]\text{-}\lim x$ if and only if A is in $(\mathcal{C}_{BP}, \mathcal{C}_{BP})$ with $BP\text{-}\lim Ax = BP\text{-}\lim_{m,n} x_{mn}$ and

$$\sum_{j,k \in E} |\Delta_{11} a_{mnjk}| \to 0 \quad (m,n \to \infty)$$

for each set E that is uniformly of zero density, where $\Delta_{11}a_{mnjk} = a_{mnjk} - a_{m,n,j+1,k} - a_{m,n,j,k+1} + a_{m,n,j+1,k+1}$ $(j,k,m,n = 0,1,\dots)$.

6 Prove that $A = (a_{mnjk}) \in ([\mathcal{F}], \mathcal{F})$ with $\mathcal{F}\text{-}\lim Ax = [\mathcal{F}]\text{-}\lim x$ if and only if A is in $(\mathcal{C}_{BP}, \mathcal{F})$ with $\mathcal{F}\text{-}\lim Ax = BP\text{-}\lim_{m,n} x_{mn}$ and

$$\sum_{j,k \in E} |\Delta_{11} a_{mnjk}| \to 0 \quad (m,n \to \infty).$$

Chapter 4
Absolute Almost Convergence and Riesz Convergence

In this chapter, we define the notion of almost bounded variation and absolute almost convergence for double sequences. We use the definition of absolute almost convergence to define absolute almost conservative and absolute almost regular matrices and find necessary and sufficient conditions to characterize these matrices.

4.1 Introduction

The space $\mathcal{BV}$ of double sequences $x = (x_{jk})$ of bounded variation was defined by Altay and Başar [4] as follows:

$$\mathcal{BV} := \left\{ x = (x_{jk}) \mid \sum_{j,k} |x_{jk} - x_{j-1,k} - x_{j,k-1} + x_{j-1,k-1}| < \infty \right\},$$

which is a Banach space normed by

$$\|x\|_{\mathcal{BV}} = \sum_{j,k} |x_{jk} - x_{j-1,k} - x_{j,k-1} + x_{j-1,k-1}|.$$

Motivated by the above definition, we define the notion of almost bounded variation. We also define absolute almost convergence for double sequences, while the idea of absolute almost convergence for single sequences was introduced in [32]. We use the definition of absolute almost convergence to define absolute almost conservative and absolute almost regular matrices and find necessary and sufficient conditions to characterize these four-dimensional matrices [96].

M. Mursaleen, S.A. Mohiuddine, *Convergence Methods for Double Sequences and Applications*, DOI 10.1007/978-81-322-1611-7_4, © Springer India 2014

4.2 Almost Bounded Variation and Absolute Almost Convergence

Let

$$\phi_{pqst}(x) = \tau_{pqst}(x) - \tau_{p-1,q,s,t}(x) - \tau_{p,q-1,s,t}(x) + \tau_{p-1,q-1,s,t}(x)$$

and

$$\tau_{pqst}(x) = \frac{1}{(p+1)(q+1)} \sum_{m=0}^{p} \sum_{n=0}^{q} x_{m+s,n+t}.$$

Thus,

$$\phi_{pqst}(x) = \frac{1}{(p+1)(q+1)} \sum_{m=0}^{p} \sum_{n=0}^{q} x_{m+s,n+t} - \frac{1}{p(q+1)} \sum_{m=0}^{p-1} \sum_{n=0}^{q} x_{m+s,n+t}$$

$$- \frac{1}{(p+1)q} \sum_{m=0}^{p} \sum_{n=0}^{q-1} x_{m+s,n+t} + \frac{1}{pq} \sum_{m=0}^{p-1} \sum_{n=0}^{q-1} x_{m+s,n+t}$$

$$= \frac{1}{(q+1)} \sum_{n=0}^{q} \left[\frac{1}{(p+1)} \sum_{m=0}^{p} x_{m+s,n+t} - \frac{1}{p} \sum_{m=0}^{p-1} x_{m+s,n+t} \right]$$

$$- \frac{1}{q} \sum_{n=0}^{q-1} \left[\frac{1}{(p+1)} \sum_{m=0}^{p} x_{m+s,n+t} - \frac{1}{p} \sum_{m=0}^{p-1} x_{m+s,n+t} \right]$$

$$= \frac{1}{(q+1)} \sum_{n=0}^{q} \left[\frac{1}{p(p+1)} \sum_{m=1}^{p} m(x_{m+s,n+t} - x_{m-1+s,n+t}) \right]$$

$$- \frac{1}{q} \sum_{n=0}^{q-1} \left[\frac{1}{p(p+1)} \sum_{m=1}^{p} m(x_{m+s,n+t} - x_{m-1+s,n+t}) \right]$$

$$= \frac{1}{p(p+1)} \sum_{m=1}^{p} m \left[\frac{1}{(q+1)} \sum_{n=0}^{q} y_{m+s,n+t} - \frac{1}{q} \sum_{n=0}^{q-1} y_{m+s,n+t} \right],$$

where $y_{m+s,n+t} = x_{m+s,n+t} - x_{m-1+s,n+t}$. Simplifying further, we get

$$\phi_{pqst}(x) = \frac{1}{p(p+1)} \sum_{m=1}^{p} m \left[\frac{1}{q(q+1)} \sum_{n=1}^{q} n(y_{m+s,n+t} - y_{m+s,n-1+t}) \right]$$

$$= \frac{1}{p(p+1)q(q+1)} \sum_{m=1}^{p} \sum_{n=1}^{q} mn[x_{m+s,n+t} - x_{m-1+s,n+t}$$

$$- x_{m+s,n-1+t} + x_{m-1+s,n-1+t}].$$

Now we write

$$
\phi_{pqst}(x) =
\begin{cases}
\frac{1}{p(p+1)q(q+1)} \sum_{m=1}^{p} \sum_{n=1}^{q} mn[x_{m+s,n+t} - x_{m-1+s,n+t} \\
\quad - x_{m+s,n-1+t} + x_{m-1+s,n-1+t}], \quad p,q \geq 1, \\
x_{st}, \quad p = 0 \text{ or } q = 0 \text{ or both } p,q = 0.
\end{cases}
\tag{4.1}
$$

Definition 4.1   A double sequence $x = (x_{jk}) \in \mathcal{M}_u$ is said to be of *almost bounded variation* if

$$
\sum_{p=0}^{\infty} \sum_{q=0}^{\infty} |\phi_{pqst}(x)| \quad \text{converges uniformly in } s, t.
$$

By $\widehat{BV}$ we denote the space of all double sequences that are of almost bounded variation.

Definition 4.2   A double sequence $x = (x_{jk}) \in \mathcal{M}_u$ is said to be *absolutely almost convergent* if

(i) $\sum_{p=0}^{\infty} \sum_{q=0}^{\infty} |\phi_{pqst}(x)|$ converges uniformly in s, t, and
(ii) $\lim_{p,q \to \infty} \tau_{pqst}(x)$, which must exist, takes the same value for all s, t.

We denote by $\hat{\mathcal{L}}$ the space of all absolutely almost convergent double sequences. It is obvious that $\hat{\mathcal{L}} \subset \widehat{BV}$.

Throughout this chapter, lim stands for *BP*-lim, and $\sum$ for *BP*-$\sum$.

Theorem 4.3 $\hat{\mathcal{L}}$ and $\widehat{BV}$ both are Banach spaces normed by

$$
\|x\| = \sup_{s,t} \sum_{p=0}^{\infty} \sum_{q=0}^{\infty} |\phi_{pqst}(x)|.
$$

Proof By uniform convergence, there exist P and Q such that

$$
\sum_{p=P+1}^{\infty} \sum_{q=Q+1}^{\infty} |\phi_{pqst}(x)| \leq 1
$$

for all s, t and for fixed P and Q,

$$
\sum_{p=0}^{P} \sum_{q=0}^{Q} |\phi_{pqst}(x)|
$$

is bounded because $x \in \hat{\mathcal{L}}$ and $\hat{\mathcal{L}} \subset \mathcal{M}_u$. Hence, $\|x\|$ is defined.

As in case of $\hat{\mathcal{L}}$ in [4], it can be easily shown that $\hat{\mathcal{L}}$ also is a normed linear space.

Now, let (x^b) be a Cauchy sequence in $\hat{\mathcal{L}}$. Then for each j, k, (x_{jk}^b) is a Cauchy sequence in $\mathbb{C}$. Therefore, $x_{jk}^b \to x_{jk}$ (say). Letting $x = (x_{jk})$, given $\epsilon > 0$, there exists an integer N such that for $b, d > N = N(\epsilon)$ and for all s, t,

$$\sum_{p=0}^{\infty} \sum_{q=0}^{\infty} \left| \phi_{pqst} \left(x^b - x^d \right) \right| < \epsilon,$$

and thus,

$$\left| \tau_{pqst} \left(x^b - x^d \right) \right| < \epsilon.$$

Taking limit $d \to \infty$, we have, for $b > N = N(\epsilon)$ and for all s, t,

$$\sum_{p=0}^{\infty} \sum_{q=0}^{\infty} \left| \phi_{pqst} \left(x^b - x \right) \right| \leq \epsilon \tag{4.2}$$

and

$$\left| \tau_{pqst} \left(x^b - x \right) \right| \leq \epsilon. \tag{4.3}$$

Now, let $\epsilon > 0$ be given. There exists a b such that (4.2) holds for all s, t. Since $x^b \in \hat{\mathcal{L}}$, we can choose p_0, q_0 such that

$$\sum_{p=p_0}^{\infty} \sum_{q=q_0}^{\infty} \left| \phi_{pqst} \left(x^b \right) \right| < \epsilon \quad \text{for all } s, t.$$

It follows from (4.2) that

$$\sum_{p=p_0}^{\infty} \sum_{q=q_0}^{\infty} \left| \phi_{pqst} \left(x^b \right) - \phi_{pqst}(x) \right| \leq \epsilon \quad \text{for all } s, t.$$

Hence

$$\sum_{p=p_0}^{\infty} \sum_{q=q_0}^{\infty} \left| \phi_{pqst}(x) \right| < 2\epsilon \quad \text{for all } s, t. \tag{4.4}$$

Thus, starting with any ϵ, we have determined p_0, q_0 such that (4.4) holds. Hence, the condition (i) of Definition 4.2 holds.

Now, for given ϵ, let (4.3) hold for fixed chosen b and for all s, t. Since $x^b \in \hat{\mathcal{L}}$, we have, for all $p \geq p_0, q \geq q_0$,

$$\left| \tau_{pqst} \left(x^b - Le \right) \right| < \epsilon \quad \text{for all } s, t.$$

It follows from (4.3) that

$$\left| \tau_{pqst} \left(x^b \right) - \tau_{pqst}(x) \right| \leq \epsilon \quad \text{for all } s, t.$$

Hence,

$$\left| \tau_{pqst}(x) - Le \right| < 2\epsilon \quad \text{for all } s, t,$$

which is condition (ii) of Definition 4.2. Hence the result. $\square$

4.3 Absolutely Almost Conservative Matrices

The idea of absolutely regular matrices for single sequences was studied by Mears [69], i.e., those matrices that transform the space v of the sequences of bounded variation into v leaving the limit invariant. Here we give the following [96].

Definition 4.4 A four-dimensional infinite matrix $A = (a_{mnjk})$ is said to be *absolutely almost conservative* if $Ax \in \hat{\mathcal{L}}$ for all $x \in \mathcal{BV}$.

Definition 4.5 An infinite matrix $A = (a_{mnjk})$ is said to be *absolutely almost regular* if and only if it is absolutely almost conservative and $\lim Ax = \lim x$ for all $x \in \mathcal{BV}$.

We write

$$\triangle_{01} x_{jk} = x_{jk} - x_{j,k-1},$$

$$\triangle_{10} x_{jk} = x_{jk} - x_{j-1,k},$$

$$\triangle_{11} x_{jk} = \triangle_{10}(\triangle_{01} x_{jk}) = \triangle_{01}(\triangle_{10} x_{jk}) = x_{jk} - x_{j-1,k} - x_{j,k-1} + x_{j-1,k-1}.$$

Now we find necessary and sufficient conditions for A to be absolutely almost conservative and absolutely almost regular.

We note that if Ax is defined, then it follows from (4.1) that, for all integers $p, q, s, t \geq 0$,

$$\sum_{j=0}^{\infty} \sum_{k=0}^{\infty} \hat{\alpha}(p, q, j, k, s, t) x_{jk},$$

where

$$\hat{\alpha}(p, q, j, k, s, t) = \begin{cases} \frac{1}{p(p+1)q(q+1)} \sum_{m=1}^{p} \sum_{n=1}^{q} mn[a_{m+s,n+t,j,k} - a_{m-1+s,n+t,j,k} \\ \qquad - a_{m+s,n-1+t,j,k} + x_{m-1+s,n-1+t,j,k}], \quad p, q \geq 1, \\ a(m, n, j, k), \quad p \text{ or } q \text{ or both zero.} \end{cases}$$

By $a(m, n, j, k)$ we denote the element a_{mnjk} of the matrix A.

Theorem 4.6 *A matrix $A = (a_{mnjk})$ is absolutely almost conservative if and only if*

(i) *there exists a constant K such that for $i, r = 0, 1, 2, \ldots$ and $s, t = 0, 1, 2, \ldots$,*

$$\sum_{p=0}^{\infty} \sum_{q=0}^{\infty} \left| \sum_{j=0}^{i} \sum_{k=0}^{r} \hat{\alpha}(p, q, j, k, s, t) \right| \leq K;$$

(ii) $u_{jk} = \sum_{p=0}^{\infty} \sum_{q=0}^{\infty} \hat{\alpha}(p, q, j, k, s, t)$ *uniformly in s, t;*

(iii) $u_{0k} = \sum_{p=0}^{\infty} \sum_{q=0}^{\infty} \sum_{j=0}^{\infty} \hat{\alpha}(p, q, j, k, s, t)$ *uniformly in* s, t;

(iv) $u_{j0} = \sum_{p=0}^{\infty} \sum_{q=0}^{\infty} \sum_{k=0}^{\infty} \hat{\alpha}(p, q, j, k, s, t)$ *uniformly in* s, t;

(v) $u = \sum_{p=0}^{\infty} \sum_{q=0}^{\infty} \sum_{j=0}^{\infty} \sum_{k=0}^{\infty} \hat{\alpha}(p, q, j, k, s, t)$ *uniformly in* s, t

$(j, k = 0, 1, 2, \ldots)$, *where* (iii), (iv), *and* (v) *respectively satisfy*

(iii)' *for each* k, $\sum_{j=0}^{\infty} a_{mnjk}$ *converges for all* m, n;

(iv)' *for each* j, $\sum_{k=0}^{\infty} a_{mnjk}$ *converges for all* m, n;

(v)' $\sum_{j=0}^{\infty} \sum_{k=0}^{\infty} a_{mnjk}$ *converges for all* m, n.

In this case, the BP-$\lim Ax$ *is*

$$u\ell - \sum_{k=0}^{\infty} u_{0k} h_k - \sum_{j=0}^{\infty} u_{j0} \ell_j + \sum_{j=0}^{\infty} \sum_{k=0}^{\infty} u_{jk} x_{jk}$$

for every $x = (x_{jk}) \in \mathcal{BV}$, *where*

$$\ell = \sum_{j=0}^{\infty} \sum_{k=0}^{\infty} \Delta_{11} x_{jk}, \qquad h_k = \sum_{j=0}^{\infty} \Delta_{10} x_{jk}, \quad \text{and} \quad \ell_j = \sum_{k=0}^{\infty} \Delta_{01} x_{jk}.$$

Theorem 4.7 *A matrix* $A = (a_{mnjk})$ *is absolutely almost regular if and only if*

(i) *there exists a constant* K *such that, for* $i, r = 0, 1, 2, \ldots$ *and* $s, t = 0, 1, 2, \ldots$,

$$\sum_{p=0}^{\infty} \sum_{q=0}^{\infty} \left| \sum_{j=0}^{i} \sum_{k=0}^{r} \hat{\alpha}(p, q, j, k, s, t) \right| \leq K;$$

(ii) $u_{jk} = 0$ *for each* j, k;

(iii) $u_{0k} = 0$ *for each* k;

(iv) $u_{j0} = 0$ *for each* j;

(v) $u = 1$.

Proof of Theorem 4.6 Let $A = (a_{mnjk})$ be absolutely almost conservative. Put

$$q_{st}(x) = \sum_{p=0}^{\infty} \sum_{q=0}^{\infty} \left| \phi_{pqst}(Ax) \right|.$$

It is clear that, for fixed s, t and for all j, k,

$$\sum_{i=0}^{j} \sum_{r=0}^{k} a_{mnir} x_{ir}$$

is a continuous linear functional on $\mathcal{BV}$. We are given that, for all $x \in \mathcal{L}$, it tends to a limit as $j, k \to \infty$ (for fixed s, t), and hence by the Banach–Steinhaus theorem, this limit, that is, $(Ax)_{st}$, is also a continuous linear functional on $\mathcal{L}$. Hence, for fixed s, t and fixed (finite) p, q,

$$\sum_{\mu=0}^{p} \sum_{\xi=0}^{q} |\phi_{\mu\xi st}(Ax)| \tag{4.5}$$

is a continuous seminorm on $\mathcal{BV}$. For any given $x \in \mathcal{BV}$, (4.5) is bounded in p, q, s, t. Hence, by another application of the Banach–Steinhaus theorem, there exists a constant $M > 0$ such that

$$q_{st}(x) \le M\|x\|. \tag{4.6}$$

Apply (4.6) with $x = (x_{jk})$ defined by

$$x_{jk} = \begin{cases} 1 & \text{if } j \le i, k \le r, \\ 0 & \text{otherwise.} \end{cases}$$

Note that, in this case, $\|x\| = 2$, and hence (i) must hold.

Since $\mathbf{e}^{jk}$, $\mathbf{e}^k$, $\mathbf{e}_j$, and $\mathbf{e}$ belong to $\mathcal{BV}$, the necessity of (ii), (iii), (iv), and (v) is obvious.

Conversely, let the conditions hold and $x = (x_{jk}) \in \mathcal{BV}$. We have defined $\hat{\mathcal{L}}$ as a subspace of $\mathcal{M}_u$. Thus, in order to prove that $Ax \in \hat{\mathcal{L}}$, it is necessary to prove that Ax exists and is bounded. Since the sum in (i) is bounded, it follows that

$$\sum_{j=0}^{i} \sum_{k=0}^{r} a_{mnjk} \tag{4.7}$$

is bounded for all i, r, m, n. Hence, by the convergence of (v)$'$ for fixed m, n, the result follows easily.

Now, by (v)$'$, the series

$$\sum_{j=0}^{\infty} \sum_{k=0}^{\infty} \hat{\alpha}(p, q, j, k, s, t)$$

converges for all p, q, s, t. Hence, if we write

$$\beta(p, q, j, k, s, t) = \sum_{i=j}^{\infty} \gamma(p, q, i, k, s, t),$$

where

$$\gamma(p, q, i, k, s, t) = \sum_{r=k}^{\infty} \hat{\alpha}(p, q, i, r, s, t),$$

then $\beta(p, q, j, k, s, t)$ is defined; also, for fixed p, q, s, t, we have

$$\left. \begin{array}{l} \gamma(p, q, i, k, s, t) \to 0 \quad \text{as } k \to \infty, \\ \beta(p, q, j, k, s, t) \to 0 \quad \text{as } j \to \infty. \end{array} \right\} \tag{4.8}$$

Then (iii) gives that

$$\sum_{p=0}^{\infty}\sum_{q=0}^{\infty}\left|\beta(p,q,0,0,s,t)\right| \tag{4.9}$$

converges uniformly in s, t. Similarly, (iii) and (iii)$'$ yield that, for fixed k,

$$\sum_{p=0}^{\infty}\sum_{q=0}^{\infty}\left|\sum_{j=0}^{\infty}\hat{\alpha}(p,q,j,k,s,t)\right| \tag{4.10}$$

converges uniformly in s, t, and (iv) and (iv)$'$ yield that, for fixed j,

$$\sum_{p=0}^{\infty}\sum_{q=0}^{\infty}\left|\sum_{k=0}^{\infty}\hat{\alpha}(p,q,j,k,s,t)\right| \tag{4.11}$$

converges uniformly in s, t. By (ii), for fixed j, k, we have that the series

$$\sum_{p=0}^{\infty}\sum_{q=0}^{\infty}\left|\hat{\alpha}(p,q,j,k,s,t)\right| \tag{4.12}$$

converges uniformly in s, t. Since

$$\beta(p,q,j,k,s,t)$$

$$=\sum_{i=j}^{\infty}\sum_{r=k}^{\infty}\hat{\alpha}(p,q,i,r,s,t)$$

$$=\left(\sum_{i=0}^{\infty}-\sum_{i=0}^{j-1}\right)\sum_{r=k}^{\infty}\hat{\alpha}(p,q,i,r,s,t)$$

$$=\left(\sum_{i=0}^{\infty}\sum_{r=k}^{\infty}-\sum_{i=0}^{j-1}\sum_{r=k}^{\infty}\right)\hat{\alpha}(p,q,i,r,s,t)$$

$$=\left[\sum_{i=0}^{\infty}\left(\sum_{r=0}^{\infty}-\sum_{r=0}^{k-1}\right)-\sum_{i=0}^{j-1}\left(\sum_{r=0}^{\infty}-\sum_{r=0}^{k-1}\right)\right]\hat{\alpha}(p,q,i,r,s,t)$$

$$=\sum_{i=0}^{\infty}\sum_{r=0}^{\infty}\hat{\alpha}(p,q,i,r,s,t)-\sum_{i=0}^{\infty}\sum_{r=0}^{k-1}\hat{\alpha}(p,q,i,r,s,t)$$

$$-\sum_{i=0}^{j-1}\sum_{r=0}^{\infty}\hat{\alpha}(p,q,i,r,s,t)+\sum_{i=0}^{j-1}\sum_{r=0}^{k-1}\hat{\alpha}(p,q,i,r,s,t)$$

$$=\beta(p,q,0,0,s,t)-\left[\beta(p,q,0,0,s,t)-\beta(p,q,0,k,s,t)\right]$$

$$- \big[\beta(p,q,0,0,s,t) - \beta(p,q,j,0,s,t)\big] + \sum_{i=0}^{j-1}\sum_{r=0}^{k-1}\hat{\alpha}(p,q,i,r,s,t)$$

$$= \beta(p,q,0,k,s,t) + \beta(p,q,j,0,s,t) - \beta(p,q,0,0,s,t)$$

$$+ \sum_{i=0}^{j-1}\sum_{r=0}^{k-1}\hat{\alpha}(p,q,i,r,s,t), \tag{4.13}$$

it follows that, for fixed j,k,

$$\sum_{p=0}^{\infty}\sum_{q=0}^{\infty}\big|\beta(p,q,j,k,s,t)\big| \tag{4.14}$$

converges uniformly in s,t.

Now

$$\phi_{pqst}(Ax) = \sum_{j=0}^{\infty}\sum_{k=0}^{\infty}\hat{\alpha}(p,q,j,k,s,t)x_{jk}$$

$$= \sum_{j=0}^{\infty}\sum_{k=0}^{\infty}\Big[\sum_{i=j}^{\infty}\sum_{r=k}^{\infty}\hat{\alpha}(p,q,i,r,s,t)\Big](\Delta_{11}x_{jk})$$

$$= \sum_{j=0}^{\infty}\sum_{k=0}^{\infty}\beta(p,q,j,k,s,t)[x_{jk} - x_{j-1,k} - x_{j,k-1} + x_{j-1,k-1}]$$

$$\tag{4.15}$$

by (4.8) and the boundedness of $x = (x_{jk})$.

Now (i) and the boundedness of the sum (4.9) show that

$$\sum_{p=0}^{\infty}\sum_{q=0}^{\infty}\big|\beta(p,q,j,k,s,t)\big| \tag{4.16}$$

is bounded for all j,k,s,t. We can make

$$\sum_{j=j_0+1}^{\infty}\sum_{k=k_0+1}^{\infty}|x_{jk} - x_{j-1,k} - x_{j,k-1} + x_{j-1,k-1}|$$

arbitrarily small by choosing j_0 and k_0 sufficiently large. Therefore, it follows that, given $\epsilon > 0$, we can choose j_0,k_0 so that, for all s,t,

$$\sum_{p=0}^{\infty}\sum_{q=0}^{\infty}\Big|\sum_{j=j_0+1}^{\infty}\sum_{k=k_0+1}^{\infty}\beta(p,q,j,k,s,t)(x_{jk} - x_{j-1,k} - x_{j,k-1} + x_{j-1,k-1})\Big| < \epsilon.$$

$$\tag{4.17}$$

Now since for each j, k, (4.14) converges uniformly in s, t, it follows that once j_0, k_0 have been chosen, we can choose p_0, q_0 so that, for all s, t,

$$\sum_{p=p_0+1}^{\infty} \sum_{q=q_0+1}^{\infty} \left| \sum_{j=0}^{j_0} \sum_{k=0}^{k_0} \beta(p,q,j,k,s,t)(x_{jk} - x_{j-1,k} - x_{j,k-1} + x_{j-1,k-1}) \right| < \epsilon.$$

It follows from (4.17) that the same inequality holds when $\sum_{p=0}^{\infty}$ and $\sum_{q=0}^{\infty}$ are replaced by $\sum_{p=p_0+1}^{\infty}$ and $\sum_{q=q_0+1}^{\infty}$, respectively; hence, for all s, t,

$$\sum_{p=p_0+1}^{\infty} \sum_{q=q_0+1}^{\infty} \left| \sum_{j=0}^{\infty} \sum_{k=0}^{\infty} \beta(p,q,j,k,s,t)(x_{jk} - x_{j-1,k} - x_{j,k-1} + x_{j-1,k-1}) \right| < 2\epsilon,$$

$$(4.18)$$

that is,

$$\sum_{p=p_0+1}^{\infty} \sum_{q=q_0+1}^{\infty} \left| \phi_{pqst}(Ax) \right| < 2\epsilon.$$

Thus,

$$\sum_{p=0}^{\infty} \sum_{q=0}^{\infty} \left| \phi_{pqst}(Ax) \right|$$

converges uniformly in s, t. Hence, Ax satisfies condition (i) of Definition 4.1; we still have to show that it satisfies condition (ii) of Definition 4.1.

If

$$\phi_{pqst}(Ax) = \sum_{j=0}^{\infty} \sum_{k=0}^{\infty} \hat{\alpha}(p,q,j,k,s,t) x_{jk},$$

then by Abel's partial sum we have

$$\phi_{pqst}(Ax) = \sum_{j=0}^{\infty} \sum_{k=0}^{\infty} \left[\sum_{i=0}^{j-1} \sum_{r=0}^{k-1} \hat{\alpha}(p,q,j,k,s,t) \right] \Delta_{11} x_{jk}.$$

Using (4.13), we get

$$\phi_{pqst}(Ax) = \sum_{j=0}^{\infty} \sum_{k=0}^{\infty} [\beta(p,q,j,k,s,t) - \beta(p,q,0,k,s,t)$$

$$- \beta(p,q,j,0,s,t) + \beta(p,q,0,0,s,t)] \Delta_{11} x_{jk}.$$

Again, using Abel's partial sum to first three summations, we get

$$\phi_{pqst}(Ax) = \sum_{j=0}^{\infty} \sum_{k=0}^{\infty} \hat{\alpha}(p,q,j,k,s,t) x_{jk} - \sum_{j=0}^{\infty} \sum_{k=0}^{\infty} \hat{\alpha}(p,q,j,k,s,t) \Delta_{10} x_{jk}$$

$$-\sum_{j=0}^{\infty}\sum_{k=0}^{\infty}\hat{\alpha}(p,q,j,k,s,t)\triangle_{01}x_{jk} + \sum_{j=0}^{\infty}\sum_{k=0}^{\infty}\hat{\alpha}(p,q,j,k,s,t)\triangle_{11}x_{jk}.$$

Hence,

$$\sum_{p=0}^{\infty}\sum_{q=0}^{\infty}\phi_{pqst}(Ax)$$

$$=\sum_{j=0}^{\infty}\sum_{k=0}^{\infty}\sum_{p=0}^{\infty}\sum_{q=0}^{\infty}\hat{\alpha}(p,q,j,k,s,t)x_{jk}$$

$$-\sum_{j=0}^{\infty}\sum_{k=0}^{\infty}\sum_{p=0}^{\infty}\sum_{q=0}^{\infty}\hat{\alpha}(p,q,j,k,s,t)\triangle_{10}x_{jk}$$

$$-\sum_{j=0}^{\infty}\sum_{k=0}^{\infty}\sum_{p=0}^{\infty}\sum_{q=0}^{\infty}\hat{\alpha}(p,q,j,k,s,t)\triangle_{01}x_{jk}$$

$$+\sum_{j=0}^{\infty}\sum_{k=0}^{\infty}\sum_{p=0}^{\infty}\sum_{q=0}^{\infty}\hat{\alpha}(p,q,j,k,s,t)\triangle_{11}x_{jk}$$

$$=\sum_{j=0}^{\infty}\sum_{k=0}^{\infty}\left[\sum_{p=0}^{\infty}\sum_{q=0}^{\infty}\hat{\alpha}(p,q,j,k,s,t)x_{jk}\right]$$

$$-\sum_{k=0}^{\infty}\left[\sum_{p=0}^{\infty}\sum_{q=0}^{\infty}\left(\sum_{j=0}^{\infty}\hat{\alpha}(p,q,j,k,s,t)(x_{jk}-x_{j-1,k})\right)\right]$$

$$-\sum_{j=0}^{\infty}\left[\sum_{p=0}^{\infty}\sum_{q=0}^{\infty}\left(\sum_{k=0}^{\infty}\hat{\alpha}(p,q,j,k,s,t)(x_{jk}-x_{j,k-1})\right)\right]$$

$$+\sum_{p=0}^{\infty}\sum_{q=0}^{\infty}\left[\sum_{j=0}^{\infty}\sum_{k=0}^{\infty}\hat{\alpha}(p,q,j,k,s,t)\right](x_{jk}-x_{j-1,k}-x_{j,k-1}+x_{j-1,k-1})$$

$$=\sum_{j=0}^{\infty}\sum_{k=0}^{\infty}u_{jk}x_{jk}-\sum_{k=0}^{\infty}u_{0k}h_k-\sum_{j=0}^{\infty}u_{j0}\ell_j+u\ell,$$

where, for $x \in \mathcal{BV}$,

$$\ell_j = \lim_{k\to\infty}x_{jk} = \sum_{k=0}^{\infty}(x_{jk}-x_{j,k-1}),$$

$$h_k = \lim_{j \to \infty} x_{jk} = \sum_{j=0}^{\infty} (x_{jk} - x_{j-1,k}),$$

$$\ell = \lim x = \sum_{j=0}^{\infty} \sum_{k=0}^{\infty} \Delta_{11} x_{jk},$$

whence the result.   □

Proof of Theorem 4.7 Suppose that A is absolutely almost regular matrix. Since e^{jk}, e^k, e_j, and $e \in \mathcal{BV}$, conditions (ii), (iii), (iv), and (v) hold, respectively. Condition (i) follows as in the proof of Theorem 4.6.

Conversely, if a matrix A satisfies the conditions of the theorem, then it is an absolutely almost conservative matrix. For $x \in \mathcal{BV}$, the *BP*-limit of Ax is

$$\sum_{j=0}^{\infty} \sum_{k=0}^{\infty} u_{jk} x_{jk} - \sum_{k=0}^{\infty} u_{0k} h_k - \sum_{j=0}^{\infty} u_{j0} \ell_j + u\ell,$$

which reduces to ℓ by using conditions (ii)–(v). Hence, A is an absolutely almost regular matrix.   □

4.4 Riesz Convergence and Matrix Transformations

In this section, we define the Riesz convergence for double sequences and determine some matrix transformations [3].

Definition 4.8 Let (q_i) and (p_j) be sequences of nonnegative numbers that are not all zero, and denote $Q_m = q_1 + q_2 + \cdots + q_m$, $q_1 > 0$, $P_n = p_1 + p_2 + \cdots + p_n$, $p_1 > 0$. Then, the transformation given by

$$t_{mn}^{qp}(x) = \frac{1}{Q_m} \frac{1}{P_n} \sum_{i=1}^{m} \sum_{j=1}^{n} q_i p_j x_{ij}$$

is called the Riesz mean of double sequence $x = (x_{jk})$. If $P\text{-}\lim t_{mn}^{qp}(x) = s$, $s \in \mathbb{R}$, then the sequence $x = (x_{jk})$ is said to be *Riesz convergent* to s.

If $x = (x_{jk})$ is Riesz convergent to s, then we write $\mathcal{R}\text{-}\lim x = s$. In what follows, $\mathcal{R}_P$ will denote the set of all Riesz convergent double sequences. Since a Riesz convergent double sequence need not be bounded, by $\mathcal{R}_{BP}$ we denote the set of all bounded and Riesz convergent double sequences. By $\mathcal{R}_{BP0}$ we denote the set of all double sequences that are bounded and Riesz convergent to zero.

Note that in the case $q_i = 1$ for all i and $p_j = 1$ for all j, the Riesz mean reduces to the Cesàro mean, and the Riesz convergence is said to be Cesàro convergence.

Now, we will give some results to characterize some classes of matrices related to the $\mathcal{R}_{BP}$.

Theorem 4.9 *A matrix* $A = (a_{mnjk}) \in (\mathcal{M}_u, \mathcal{R}_{BP0})$ *if and only if*

$$\|A\| = \sup_{m,n} \sum_{j,k} |a_{mnjk}| < \infty, \tag{4.19}$$

$$P\text{-}\lim_{m,n} \eta(m,n,r,s,q,p) = 0 \quad (r,s \in \mathbb{N}), \tag{4.20}$$

$$P\text{-}\lim_{m,n} \sum_r |\eta(m,n,r,s,q,p)| = 0 \quad (s \in \mathbb{N}), \tag{4.21}$$

$$P\text{-}\lim_{m,n} \sum_s |\eta(m,n,r,s,q,p)| = 0 \quad (r \in \mathbb{N}), \tag{4.22}$$

$$P\text{-}\lim_{m,n} \sum_{r,s} |\eta(m,n,r,s,q,p)| = 0, \tag{4.23}$$

where

$$\eta(m,n,r,s,q,p) = \frac{1}{Q_r}\frac{1}{P_s}\sum_{j=1}^{r}\sum_{k=1}^{s} q_j p_k a_{mnjk}.$$

Proof Let $A = (a_{mnjk}) \in (\mathcal{M}_u, \mathcal{R}_{BP0})$. This means that Ax exists for all $x = (x_{jk}) \in \mathcal{M}_u$ and $Ax \in \mathcal{R}_{BP0}$, which implies (4.19). Let us define the sequence $y = (y_{rs})$ by

$$y_{rs} = \begin{cases} \operatorname{sgn} \eta(m_i, n_j, r, s, q, p), & r_{i-1} < r < r_i, s_{j-1} < s < s_j, \\ 0 & \text{otherwise.} \end{cases}$$

Then, the necessity of (4.23) follows from $P\text{-}\lim t_{rs}^{qp}(Ax)$.

It is known by the assumption that

$$P\text{-}\lim \sum_{r,s} \eta(m,n,r,s,q,p) x_{jk} = 0.$$

So, if we define the sequences e_{ij}^{rs}, e_r, e^s by

$$e_{ij}^{rs} = \begin{cases} 1, & (j,k) = (r,s), \\ 0, & \text{otherwise,} \end{cases}$$

$e_r = \sum_s e^{rs}$ $(r \in \mathbb{N})$, and $e^s = \sum_r e^{rs}$ $(s \in \mathbb{N})$, then the necessity of (4.20), (4.21), and (4.22) follows from $P\text{-}\lim t_{rs}^{qp}(Ae_{rs})$, $P\text{-}\lim t_{rs}^{qp}(Ae_r)$, and $P\text{-}\lim t_{rs}^{qp}(Ae^s)$, respectively.

Since the proof of the converse part is routine, we omit the details. $\quad\square$

Theorem 4.10 *A matrix* $A = (a_{mnjk}) \in (\mathcal{C}_{BP}, \mathcal{R}_{BP})_{reg}$ *if and only if* (4.19)–(4.22) *hold and*

$$P\text{-}\lim_{m,n} \sum_{r,s} |\eta(m, n, r, s, q, p)| = 1. \qquad (4.24)$$

Proof The necessity of the conditions can be shown by the same way used in the proof of Theorem 4.9.

For the sufficiency, let the conditions hold, and $x = (x_{jk}) \in \mathcal{C}_{BP}$ with $P\text{-}\lim x_{jk} = L$ (say). Then, there exists $N > 0$ such that $|x_{jk}| < |L| + \varepsilon$ for all $j, k > N$. Now, let us write

$$\sum_{r,s} \eta(m, n, r, s, q, p) x_{rs}$$

$$= \sum_{r=0}^{N} \sum_{s=0}^{N} \eta(m, n, r, s, q, p) x_{rs} + \sum_{r=N+1}^{\infty} \sum_{s=0}^{N-1} \eta(m, n, r, s, q, p) x_{rs}$$

$$+ \sum_{r=0}^{N-1} \sum_{s=N+1}^{\infty} \eta(m, n, r, s, q, p) x_{rs} + \sum_{r=N+1}^{\infty} \sum_{s=N+1}^{\infty} \eta(m, n, r, s, q, p) x_{rs},$$

which implies that

$$\left| \sum_{r,s} \eta(m, n, r, s, q, p) x_{rs} \right|$$

$$= \|x\| \sum_{r=0}^{N} \sum_{s=0}^{N} |\eta(m, n, r, s, q, p)| + \|x\| \sum_{r=N+1}^{\infty} \sum_{s=0}^{N-1} |\eta(m, n, r, s, q, p)|$$

$$+ \|x\| \sum_{r=0}^{N-1} \sum_{s=N+1}^{\infty} |\eta(m, n, r, s, q, p)|$$

$$+ (|L| + \varepsilon) \left| \sum_{r=N+1}^{\infty} \sum_{s=N+1}^{\infty} \eta(m, n, r, s, q, p) \right|.$$

So, by letting $m, n \to \infty$ under the light of the assumption, we get that $P\text{-}\lim t_{rs}^{qp}(Ax) = L$. This completes the proof. □

4.5 Exercises

1 Characterize the class $(\mathcal{L}_u, \hat{\mathcal{L}})$.

2 Find necessary and sufficient conditions for the matrix class $(\mathcal{BV}, \widehat{\mathcal{BV}})$.

3 Find necessary and sufficient conditions for the matrix class $(\mathcal{M}_u, \hat{\mathcal{L}})$.

4 Find necessary and sufficient conditions for the matrix class $(\mathcal{M}_u, \widehat{\mathcal{BV}})$.

5 Characterize the matrix class $(\mathcal{F}, \hat{\mathcal{L}})$.

6 State and prove conditions to characterize the matrix class $(\mathcal{F}, \widehat{\mathcal{BV}})$.

7 State and prove conditions to characterize the matrix class $(\mathcal{F}, \mathcal{R}_{BP})_{\text{reg}}$.

Chapter 5
Almost Convergence and Core Theorems

In this chapter, we define P-core, M-core, and R-cores by using the notion of Pringsheim's convergence, almost convergence, and Riesz's convergence of double sequences. We present various core theorems analogous to the well-known Knopp core theorem.

5.1 Introduction

A two-dimensional matrix transformation is said to be *regular* if it maps every convergent sequence into a convergent sequence with the same limit. In 1926, Robinson [111] presented a four-dimensional analogue of regularity for double sequences, in which he added an additional assumption of boundedness.

The following is a four-dimensional analogue of the well-known Silverman–Toeplitz theorem [25].

Theorem 5.1 *The four-dimensional matrix* $A = (a_{mnjk})$ *is bounded-regular or RH-regular if and only if (see Hamilton [50] and Robinson [111])*

(RH$_1$) $BP\text{-}\lim_{m,n} a_{mnjk} = 0$ $(j, k = 0, 1, \ldots)$,
(RH$_2$) $BP\text{-}\lim_{m,n} \sum_{j,k=0,0}^{\infty,\infty} a_{mnjk} = 1$,
(RH$_3$) $BP\text{-}\lim_{m,n} \sum_{j=0}^{\infty} |a_{mnjk}| = 0$ $(k = 0, 1, \ldots)$,
(RH$_4$) $BP\text{-}\lim_{m,n} \sum_{k=0}^{\infty} |a_{mnjk}| = 0$ $(j = 0, 1, \ldots)$,
(RH$_5$) $\sum_{j,k=0,0}^{\infty,\infty} |a_{mnjk}| \leq C < \infty$ $(m, n = 0, 1, \ldots)$.

Note that (RH$_1$) *is a consequence of each of* (RH$_3$) *and* (RH$_4$).

In Chap. 3, we have defined and characterized the almost $\mathcal{C}_{BP}$-regular matrices (Theorem 3.4). For our convenience, we call these four-dimensional matrices simply as almost regular matrices. That is, a matrix $A = (a_{mnjk})$ is almost regular if it transforms bounded convergent sequences into almost convergent sequences with

M. Mursaleen, S.A. Mohiuddine, *Convergence Methods for Double Sequences and Applications*, DOI 10.1007/978-81-322-1611-7_5, © Springer India 2014

$\mathcal{F}$-$\lim Ax = BP$-$\lim x$, for which necessary and sufficient conditions are as follows.

Theorem 5.2 *A matrix $A = (a_{mnjk})$ is almost regular if and only if the following conditions hold:*

(i) $\sup_{m,n} \sum_{j,k} |a_{mnjk}| =: M < \infty,$

(ii) *the limit BP-$\lim_{p,q} \alpha(j, k, p, q, s, t) = 0$ exists $(j, k \in \mathbb{N})$ uniformly in $s, t \in \mathbb{N}$,*

(iii) *the limit BP-$\lim_{p,q} \sum_{j,k} \alpha(j, k, p, q, s, t) = 1$ exists uniformly in $s, t \in \mathbb{N}$,*

(iv) *the limit BP-$\lim_{p,q} \sum_k |\alpha(j, k, p, q, s, t)| = 0$ exists $(j \in \mathbb{N})$ uniformly in $s, t \in \mathbb{N}$,*

(v) *the limit BP-$\lim_{p,q} \sum_j |\alpha(j, k, p, q, s, t)| = 0$ exists $(k \in \mathbb{N})$ uniformly in $s, t \in \mathbb{N}$,*

where

$$\alpha(j, k, p, q, s, t) = \frac{1}{pq} \sum_{m=s}^{s+p-1} \sum_{n=t}^{t+q-1} a_{mnjk}.$$

In this section, we prove some core theorems for double sequences using the notion of P-convergence and almost convergence. Results of this chapter are borrowed from [3, 88, 94, 103], and [107].

5.2 Pringsheim's Core

In [107], Patterson extended this idea for double sequences by defining the Pringsheim's core as follows.

Let P-$C_n\{x\}$ be the least closed convex set that includes all points x_{jk} for $j, k > n$; then the *Pringsheim's core (or P-core)* of the double sequence $x = (x_{jk})$ is the set

$$P\text{-}C\{x\} = \bigcap_{n=1}^{\infty} [P\text{-}C_n\{x\}]. \tag{5.1}$$

Note that the Pringsheim core of a real-valued bounded double sequence is the closed interval $[P$-$\liminf x, P$-$\limsup x]$.

Lemma 5.3 *If $A = (a_{mnjk})$ is a real or complex-valued four-dimensional matrix such that (RH3), (RH4), and*

$$P\text{-}\limsup_{m,n} \sum_{j,k=0,0}^{\infty,\infty} |a_{mnjk}| = M$$

hold, then for any bounded double sequence $x = (x_{jk})$, we have

$$P\text{-}\limsup |y| \leq M\big(P\text{-}\limsup |x|\big),$$

where

$$y_{mn} = \sum_{j,k=0,0}^{\infty,\infty} a_{mnjk} x_{jk}.$$

Proof Let $x = (x_{jk})$ be a bounded double sequence and define

$$B := P\text{-}\limsup x < \infty.$$

Given $\epsilon > 0$, we can choose an N such that $|x_{jk}| < (B + \epsilon)/3$ for each j and/or $k > N$. Thus,

$$
\begin{aligned}
y_{mn} \leq{}& \sum_{j,k=0,0}^{N,N} |a_{mnjk}||x_{jk}| + \sum_{\substack{0 \leq k \leq N \\ N < j < \infty}} |a_{mnjk}||x_{jk}| \\
&+ \sum_{\substack{N < k \leq \infty \\ 0 \leq j \leq N}} |a_{mnjk}||x_{jk}| + \sum_{j,k=N+1,N+1}^{\infty,\infty} |a_{mnjk}||x_{jk}| \\
\leq{}& \sum_{j,k=0,0}^{N,N} |a_{mnjk}||x_{jk}| + \sum_{\substack{0 \leq k \leq N \\ N < j < \infty}} |a_{mnjk}|\left(\frac{B+\epsilon}{3}\right) \\
&+ \sum_{\substack{N < k \leq \infty \\ 0 \leq j \leq N}} |a_{mnjk}|\left(\frac{B+\epsilon}{3}\right) + \sum_{j,k=N+1,N+1}^{\infty,\infty} |a_{mnjk}|\left(\frac{B+\epsilon}{3}\right),
\end{aligned}
$$

which yields

$$P\text{-}\limsup |y| \leq M(B + \epsilon).$$

Thus, we have

$$P\text{-}\limsup |y| \leq M\big(P\text{-}\limsup |x|\big).$$

Since

$$P\text{-}\limsup_{m,n} \sum_{j,k=0,0}^{\infty,\infty} |a_{mnjk}| = M,$$

we may assume that $M > 0$ without loss of generality. Using the RH-regularity conditions, we choose m_0, n_0, j_0, and k_0 so large that

$$\sum_{j,k=0,0}^{\infty,\infty} |a_{m_0 n_0 jk}| > M - \frac{1}{4},$$

$$\sum_{\substack{0 < k < k_0 \\ j_0 \le j \le \infty}} |a_{m_0 n_0 jk}| \le \frac{1}{4},$$

$$\sum_{\substack{k_0 \le k \le \infty \\ 0 < j < j_0}} |a_{m_0 n_0 jk}| \le \frac{1}{4},$$

and

$$\sum_{j,k=j_0,k_0}^{\infty,\infty} |a_{m_0 n_0 jk}| \le \frac{1}{4}.$$

Let $[m_p - 1]$, $[n_q - 1]$, $[j_p - 1]$, and $[l_q - 1]$ be four chosen strictly increasing index sequences with $p, q = 1, \ldots, s - 1, t - 1$ with $j_0 = k_0 > 0$. Using the RH-regularity conditions, we now choose $m_s > m_{s-1}$ and $n_t > n_{t-1}$ such that

$$\sum_{\substack{0 \le j \le j_{s-1} \\ 0 \le k \le \infty}} |a_{m_s n_t jk}| < \frac{1}{2^{s+t}},$$

$$\sum_{\substack{0 \le k \le k_{t-1} \\ j_{s-1} < j \le \infty}} |a_{m_s n_t jk}| < \frac{1}{2^{s+t}},$$

and

$$\sum_{j,k=0,0}^{\infty,\infty} |a_{m_s n_t jk}| > M - \frac{1}{2^{s+t}}.$$

In addition, we also choose $j_s > j_{s-1}$ and $k_t > k_{t-1}$ such that

$$\sum_{\substack{j_{s-1} < j < j_s \\ k_t \le k \le \infty}} |a_{m_s n_t jk}| < \frac{1}{2^{s+t}}$$

and

$$\sum_{\substack{k_{t-1} < k < \infty \\ j_s \le j \le \infty}} |a_{m_s n_t jk}| < \frac{1}{2^{s+t}}.$$

Let us define a double sequence $x = (x_{jk})$ as follows:

$$x_{jk} = \begin{cases} \dfrac{\bar{a}_{m_s n_t jk}}{|a_{m_s n_t jk}|} & \text{if } j_{s-1} < j < j_s,\ k_{t-1} < k < k_t,\ \text{and } a_{m_s n_t jk} \neq 0, \\ 0 & \text{otherwise.} \end{cases}$$

Now,

$$|y_{m_s n_t}| = \left| \sum_{j,k=0,0}^{\infty,\infty} a_{m_s n_t jk} x_{jk} \right|$$

$$\geq - \sum_{\substack{0 \leq j \leq j_s-1 \\ 0 \leq k \leq \infty}} |a_{m_s n_t jk}| - \sum_{\substack{0 \leq k \leq k_t-1 \\ j_{s-1} < j \leq \infty}} |a_{m_s n_t jk}| - \sum_{\substack{j_{s-1} < j < j_s \\ k_t \leq k \leq \infty}} |a_{m_s n_t jk}|$$

$$- \sum_{\substack{k_t-1 < k < \infty \\ js \leq j \leq \infty}} |a_{m_s n_t jk}| + \sum_{\substack{k_{t-1} < k < k_t \\ j_{s-1} < j < j_t}} a_{m_s n_t jk}\, \text{sgn}(a_{m_s n_t jk})$$

$$\geq - \frac{1}{2^{s+t}} - \frac{1}{2^{s+t}} - \frac{1}{2^{s+t}} - \frac{1}{2^{s+t}} + M - 5\left(\frac{1}{2^{s+t}}\right)$$

$$= M - 9\left(\frac{1}{2^{s+t}}\right).$$

This implies that

$$P\text{-}\limsup |y| \geq M = M\big(P\text{-}\limsup |x|\big). \qquad \square$$

Theorem 5.4 *If A is a four-dimensional matrix, then for all real-valued bounded double sequences $x = (x_{jk})$,*

$$P\text{-}\limsup Ax \leq P\text{-}\limsup x$$

if and only if

(i) *A is RH-regular, and*
(ii) *$P\text{-}\lim_{m,n} \sum_{j,k=0,0}^{\infty,\infty} |a_{mnjk}| = 1$.*

Proof (Necessity.) Let $x = (x_{jk})$ be a bounded P-convergent double sequence. Thus,

$$P\text{-}\liminf x = P\text{-}\limsup x = P\text{-}\lim x,$$

and also

$$P\text{-}\limsup A(-x) \leq -(P\text{-}\liminf x).$$

This implies that $P\text{-}\liminf x \leq P\text{-}\liminf Ax$. Therefore,

$$P\text{-}\liminf x \leq P\text{-}\liminf Ax \leq P\text{-}\limsup Ax \leq P\text{-}\limsup x.$$

Hence, Ax is P-convergent, and $P\text{-}\lim Ax = P\text{-}\lim x$. Thus, A is an RH-regular summability matrix.

By Lemma 5.3 there exists a bounded double sequence $x = (x_{jk})$ such that $P\text{-}\limsup |x| = 1$ and $P\text{-}\limsup y = C$, where C is defined by (RH$_5$). This implies that

$$1 \le P\text{-}\liminf_{m,n} \sum_{j,k=0,0}^{\infty,\infty} |a_{mnjk}| \le P\text{-}\limsup_{m,n} \sum_{j,k=0,0}^{\infty,\infty} |a_{mnjk}| \le 1,$$

whence

$$P\text{-}\lim_{m,n} \sum_{j,k=0,0}^{\infty,\infty} |a_{mnjk}| = 1.$$

(*Sufficiency.*) For $p, q > 1$, we obtain:

$$\left| \sum_{j,k=0,0}^{\infty,\infty} a_{mnjk} x_{jk} \right|$$

$$= \left| \sum_{j,k=0,0}^{\infty,\infty} \frac{|a_{mnjk}x_{jk}| - a_{mnjk}x_{jk}}{2} + \sum_{j,k=0,0}^{\infty,\infty} \frac{|a_{mnjk}x_{jk}| + a_{mnjk}x_{jk}}{2} \right|$$

$$\le \sum_{j,k=0,0}^{\infty,\infty} |a_{mnjk}||x_{jk}| + \sum_{j,k=0,0}^{\infty,\infty} \left(|a_{mnjk}| - a_{mnjk} \right) |x_{jk}|$$

$$\le \|x\| \sum_{j,k=0,0}^{p,q} |a_{mnjk}| + \|x\| \sum_{\substack{p < j < \infty \\ 0 \le k \le q}} |a_{mnjk}| + \|x\| \sum_{\substack{0 \le j < p \\ q < k < \infty}} |a_{mnjk}|$$

$$+ \sup_{j,k>p,q} |x| \sum_{j,k>p,q} |a_{mnjk}| + \|x\| \sum_{j,k=0,0}^{\infty,\infty} \left(|a_{mnjk}| - a_{mnjk} \right).$$

Using (RH$_1$)–(RH$_4$) and

$$P\text{-}\lim_{m,n} \sum_{j,k=0,0}^{\infty,\infty} |a_{mnjk}| = 1,$$

we take Pringsheim limits and get the desired result. $\square$

5.3 M-Core

In this section, we define the M-core of a double sequence by using the idea of almost convergence of double sequences and prove some core theorems by using M-core.

Let us write

$$L^{\star}(x) = P\text{-}\limsup_{p,q\to\infty}\sup_{m,n\geq 0}\frac{1}{pq}\sum_{j=m}^{m+p-1}\sum_{k=n}^{n+q-1}x_{jk}. \tag{5.2}$$

Then we define the *M*-core of a real bounded double sequence $x = (x_{jk})$ to be the closed interval $[-L^{\star}(-x), L^{\star}(x)]$.

Since every bounded convergent double sequence is almost convergent, we have

$$L^{\star}(x) \leq P\text{-}\limsup x,$$

and hence it follows that *M*-core$\{x\} \subseteq P$-core$\{x\}$ for a bounded double sequence $x = (x_{jk})$.

Here we prove some core theorems for *M*-core by applying the almost regular, strongly regular, and almost strongly regular four-dimensional matrices.

Theorem 5.5 *For every real bounded double sequence* $x = (x_{jk})$,

$$L^{\star}(Ax) \leq L(x) \tag{5.3}$$

(or M-core$\{Ax\} \subseteq P$-core$\{x\}$*) if and only if*

(i) $A = (a_{mnjk})$ *is almost regular, and*
(ii) $P\text{-}\lim_{p,q\to\infty}\sum_{j,k=0,0}^{\infty,\infty}|\alpha(j,k,p,q,s,t)| = 1$, *uniformly in* s, t.

Proof (*Necessity.*) Let us consider a real bounded double sequence x convergent to some real number λ. Then by (5.3) we get

$$\lambda = -L(-x) \leq -L^{\star}(-Ax) \leq L^{\star}(Ax) \leq L(x) = \lambda.$$

Hence, Ax is almost convergent, and $\mathcal{F}\text{-}\lim Ax = BP\text{-}\lim x = \lambda$, and so A is almost regular. Now by the above Lemma 5.3 there exists $x \in \mathcal{M}_u$ such that $\|x\| \leq 1$ and

$$L^{\star}(Ax) = P\text{-}\limsup_{p,q\to\infty}\sup_{s,t\geq 0}\sum_{j,k=0,0}^{\infty,\infty}|\alpha(j,k,p,q,s,t)| = 1.$$

Hence, if we define $x = (x_{jk})$ by

$$x_{jk} = \begin{cases} 1 & \text{if } j = k, \\ 0 & \text{otherwise}, \end{cases}$$

then

$$1 = l^{\star}(Ax) = P\text{-}\liminf_{p,q\to\infty}\sup_{s,t\geq 0}\sum_{j,k=0,0}^{\infty,\infty}|\alpha(j,k,p,q,s,t)|$$

$$\leq L^{\star}(Ax) \leq L(x) \leq \|x\| \leq 1,$$

and hence (ii) is satisfied, where

$$l^{\star}(x) = P\text{-}\liminf_{p,q\to\infty}\ \sup_{m,n\geq 0}\ \frac{1}{pq}\sum_{j=m}^{m+p-1}\sum_{k=n}^{n+q-1} x_{jk}.$$

(*Sufficiency.*) For $M, N > 1$, we obtain

$$\left|\sum_{j=0}^{\infty}\sum_{k=0}^{\infty}\frac{1}{pq}\sum_{m=s}^{s+p-1}\sum_{n=t}^{t+q-1} a_{mnjk}x_{jk}\right|$$

$$= \left|\sum_{j=0}^{\infty}\sum_{k=0}^{\infty}\left(\frac{|\alpha(j,k,p,q,s,t)|+\alpha(j,k,p,q,s,t)}{2}\right.\right.$$

$$\left.\left. -\frac{|\alpha(j,k,p,q,s,t)|-\alpha(j,k,p,q,s,t)}{2}\right)x_{jk}\right|$$

$$\leq \sum_{j=0}^{\infty}\sum_{k=0}^{\infty}|\alpha(j,k,p,q,s,t)||x_{jk}|$$

$$+\sum_{j=0}^{\infty}\sum_{k=0}^{\infty}|(|\alpha(j,k,p,q,s,t)|-\alpha(j,k,p,q,s,t))x_{jk}|$$

$$\leq \|x\|\sum_{j=0}^{M}\sum_{k=0}^{N}|\alpha(j,k,p,q,s,t)|+\|x\|\sum_{j=M+1}^{\infty}\sum_{k=0}^{N}|\alpha(j,k,p,q,s,t)|$$

$$+\|x\|\sum_{j=0}^{M}\sum_{k=N+1}^{\infty}|\alpha(j,k,p,q,s,t)|$$

$$+\left(\sup_{j,k\geq M,N}|x_{jk}|\right)\sum_{j=M+1}^{\infty}\sum_{k=N+1}^{\infty}|\alpha(j,k,p,q,s,t)|$$

$$+\|x\|\sum_{j=0}^{\infty}\sum_{k=0}^{\infty}(|\alpha(j,k,p,q,s,t)|-\alpha(j,k,p,q,s,t)).$$

Using the conditions of almost regularity and condition (ii), we get

$$L^{\star}(Ax) \leq L(x). \qquad\qquad \square$$

Theorem 5.6 *For every bounded double sequence* $x = (x_{jk})$,

$$L(Ax) \leq L^{\star}(x) \tag{5.4}$$

(*or* P-core$\{Ax\} \subseteq$ M-core$\{x\}$) *if and only if*

(i) $A = (a_{mnjk})$ is strongly regular, and

(ii) $P\text{-}\lim_{m,n\to\infty} \sum_{j,k=0,0}^{\infty,\infty} |a_{mnjk}| = 1$.

Proof (*Necessity.*) Let us consider a bounded double sequence x to be almost convergent to s. Then we have $L^*(x) = -L^*(-x)$. By (5.4) we get

$$s = -L^*(-x) \le -L(-Ax) \le L(Ax) \le L^*(x) = s.$$

Hence, Ax is BP-convergent, and $BP\text{-}\lim Ax = \mathcal{F}\text{-}\lim x = s$, and so A is strongly regular, i.e., condition (i) holds. Since every strongly regular matrix is also bounded-regular, by Lemma 5.3 there exists a bounded double sequence $x = (x_{jk})$ such that $P\text{-}\limsup |x| = 1$ and $P\text{-}\limsup |Ax| = C$, where C is defined by (RH5). Therefore, we have

$$1 \le P\text{-}\liminf_{m,n} \sum_{j,k=0,0}^{\infty,\infty} |a_{mnjk}| \le P\text{-}\limsup_{m,n} \sum_{j,k=0,0}^{\infty,\infty} |a_{mnjk}| \le 1,$$

i.e., condition (ii) holds.

(*Sufficiency.*) Given $\epsilon > 0$, we can find fixed integers $p, q \ge 2$ such that

$$\frac{1}{pq} \sum_{j=m}^{m+p-1} \sum_{k=n}^{n+q-1} x_{jk} < L^*(x) + \epsilon. \tag{5.5}$$

Now, as in [83], we can write

$$y_{MN} = \sum_{j,k=0,0}^{\infty,\infty} a_{MNjk} x_{jk} = \Sigma_1 + \Sigma_2 + \Sigma_3 + \Sigma_4 + \Sigma_5 + \Sigma_6 + \Sigma_7 + \Sigma_8, \tag{5.6}$$

where

$$\Sigma_1 = \frac{1}{pq} \sum_{m,n=0,0}^{\infty,\infty} a_{MNmn} \sum_{j=m}^{m+p-1} \sum_{k=n}^{n+q-1} x_{jk},$$

$$\Sigma_2 = \frac{-1}{pq} \sum_{j=0}^{p-2} \sum_{k=0}^{q-2} x_{jk} \sum_{m=0}^{j} \sum_{n=0}^{k} a_{MNmn},$$

$$\Sigma_3 = -\frac{1}{pq} \sum_{j=p-1}^{\infty} \sum_{k=0}^{q-2} x_{jk} \sum_{m=j-p+1}^{j} \sum_{n=0}^{k} a_{MNmn},$$

$$\Sigma_4 = -\frac{1}{pq} \sum_{j=0}^{p-2} \sum_{k=q-1}^{\infty} x_{jk} \sum_{m=0}^{j} \sum_{n=k-q+1}^{k} a_{MNmn},$$

$$\Sigma_5 = -\sum_{j=p-1}^{\infty} \sum_{k=q-1}^{\infty} x_{jk} \left\{ \frac{1}{pq} \sum_{m=j-p+1}^{j} \sum_{n=k-q+1}^{k} a_{MNmn} - a_{MNjk} \right\},$$

$$\Sigma_6 = \sum_{j=0}^{p-2} \sum_{k=0}^{q-2} a_{MNjk} x_{jk},$$

$$\Sigma_7 = \sum_{j=p-1}^{\infty} \sum_{k=0}^{q-2} a_{MNjk} x_{jk},$$

$$\Sigma_8 = -\sum_{j=0}^{p-2} \sum_{k=q-1}^{\infty} a_{MNjk} x_{jk}.$$

Using the conditions of strong regularity of A, we observe that

$$|\Sigma_2| \le \|x\| \sum_{m=0}^{p-2} \sum_{n=0}^{q-2} |a_{MNmn}| \to 0 \quad (M, N \to \infty)$$

and

$$|\Sigma_6| \le \|x\| \sum_{j=0}^{p-2} \sum_{k=0}^{q-2} |a_{MNjk}| \to 0, \quad \text{by (RH}_1\text{)};$$

$$|\Sigma_3| \le \|x\| \sum_{m=0}^{\infty} \sum_{n=0}^{q-2} |a_{MNmn}| \to 0$$

and

$$|\Sigma_7| \le \|x\| \sum_{j=p-1}^{\infty} \sum_{k=0}^{q-2} |a_{MNjk}| \to 0, \quad \text{by (RH}_3\text{)};$$

$$|\Sigma_4| \to 0 \quad \text{and} \quad |\Sigma_8| \to 0 \quad \text{by (RH}_4\text{)}.$$

Now

$$|\Sigma_5| \le \frac{\|x\|}{pq} \sum_{r=0}^{p-1} \sum_{s=0}^{q-1} \left\{ (p-r-1) \sum_{j=0}^{\infty} \sum_{k=0}^{\infty} |\Delta_{10} a_{MNjk}| \right.$$

$$\left. + (q-s-1) \sum_{j=0}^{\infty} \sum_{k=0}^{\infty} |\Delta_{01} a_{MNjk}| \right\} \to 0 \quad \text{by (3.4) and (3.5) of Chap. 3.}$$

Therefore, by (5.6) we have

$$L(Ax) \leq P\text{-}\limsup_{M,N} \sum_{m,n=0,0}^{\infty,\infty} a_{MNmn} \frac{1}{pq} \sum_{j=m}^{m+p-1} \sum_{k=n}^{n+q-1} x_{jk}$$

$$\leq P\text{-}\limsup_{M,N} \left| \sum_{m,n=0,0}^{\infty,\infty} \left(\frac{|a_{MNmn}| + a_{MNmn}}{2} + \frac{|a_{MNmn}| - a_{MNmn}}{2} \right) \frac{1}{pq} \right.$$

$$\times \left. \sum_{j=m}^{m+p-1} \sum_{k=n}^{n+q-1} x_{jk} \right|$$

$$\leq P\text{-}\limsup_{M,N} \left\{ \sum_{m,n=0,0}^{\infty,\infty} |a_{MNmn}| \left| \frac{1}{pq} \sum_{j=m}^{m+p-1} \sum_{k=n}^{n+q-1} x_{jk} \right| \right.$$

$$+ \|x\| \sum_{m,n=0,0}^{\infty,\infty} \left(|a_{MNmn}| - a_{MNmn} \right) \Bigg\}.$$

Now conditions (RH$_1$), (RH$_5$), and (ii) yield

$$L(Ax) \leq L^{\star}(x) + \epsilon.$$

Since ϵ is arbitrary, we finally have

$$L(Ax) \leq L^{\star}(x). \qquad \qquad \square$$

Now, we use the concept of almost strongly regular matrices to establish the following core theorem.

Theorem 5.7 *For every real bounded double sequence* x,

$$L^{\star}(Ax) \leq L^{\star}(x) \qquad\qquad (5.7)$$

(or M-*core$\{Ax\} \subseteq M$-core$\{x\}$) if and only if*

(i) $A = (a_{mnjk})$ *is almost strongly regular, and*
(ii) $P\text{-}\limsup_{p,q\to\infty} \sum_{j,k=0,0}^{\infty,\infty} |\alpha(j,k,p,q,s,t)| = 1$ *uniformly in* $s,t = 1,2,\ldots$.

Proof (Necessity.) Let us consider a real bounded double sequence x almost convergent to some real number λ. Then by (5.7) we get

$$\lambda = -L^{\star}(-x) \leq -L^{\star}(-Ax) \leq L^{\star}(Ax) \leq L^{\star}(x) = \lambda.$$

Hence, Ax is almost convergent, and $\mathcal{F}\text{-}\lim Ax = \mathcal{F}\text{-}\lim x = \lambda$, and so A is almost strongly regular. Now by Lemma 5.3, there exists $x \in \mathcal{M}_u$ such that $\|x\| \leq 1$ and

$$L^{\star}(Ax) = P\text{-}\limsup_{p,q\to\infty} \sup_{s,t\geq 0} \sum_{j,k=0,0}^{\infty,\infty} |\alpha(j,k,p,q,s,t)|.$$

Hence, if we define $x = (x_{jk})$ by

$$x_{jk} = \begin{cases} 1 & \text{if } j=k, \\ 0 & \text{otherwise,} \end{cases}$$

then

$$1 = l^{\star}(Ax) = P\text{-}\liminf_{p,q\to\infty} \sup_{s,t\geq 0} \sum_{j,k=0,0}^{\infty,\infty} |\alpha(j,k,p,q,s,t)|$$

$$\leq L^{\star}(Ax) \leq L^{\star}(x) \leq \|x\| \leq 1,$$

and hence (ii) is satisfied, where

$$l^{\star}(x) = P\text{-}\liminf_{p,q\to\infty} \sup_{m,n\geq 0} \frac{1}{pq} \sum_{j=m}^{m+p-1} \sum_{k=n}^{n+q-1} x_{jk}.$$

(*Sufficiency.*) Given $\epsilon > 0$, we can find fixed integers $p, q \geq 2$ such that

$$\frac{1}{pq} \sum_{j=m}^{m+p-1} \sum_{k=n}^{n+q-1} x_{jk} < L^{\star}(x) + \epsilon. \qquad (5.8)$$

For $M, N > 1$, we obtain

$$\sum_{j=0}^{\infty} \sum_{k=0}^{\infty} \frac{1}{pq} \sum_{m=s}^{s+p-1} \sum_{n=t}^{t+q-1} a_{mnjk} x_{jk}$$

$$= \sum_{j=0}^{\infty} \sum_{k=0}^{\infty} \left(\frac{|\alpha(j,k,p,q,s,t)| + \alpha(j,k,p,q,s,t)}{2} \right.$$

$$\left. - \frac{|\alpha(j,k,p,q,s,t)| - \alpha(j,k,p,q,s,t)}{2} \right) x_{jk}$$

$$\leq \sum_{j=0}^{\infty} \sum_{k=0}^{\infty} |\alpha(j,k,p,q,s,t)| x_{jk}$$

$$+ \sum_{j=0}^{\infty} \sum_{k=0}^{\infty} (|\alpha(j,k,p,q,s,t)| - \alpha(j,k,p,q,s,t)) x_{jk}$$

$$\leq \|x\| \sum_{j=0}^{M} \sum_{k=0}^{N} |\alpha(j,k,p,q,s,t)| + \|x\| \sum_{j=M+1}^{\infty} \sum_{k=0}^{N} |\alpha(j,k,p,q,s,t)|$$

$$+ \|x\| \sum_{j=0}^{M} \sum_{k=N+1}^{\infty} |\alpha(j,k,p,q,s,t)| + \sum_{j=M+1}^{\infty} \sum_{k=N+1}^{\infty} |\alpha(j,k,p,q,s,t)| x_{jk}$$

$$+ \|x\| \sum_{j=0}^{\infty} \sum_{k=0}^{\infty} (|\alpha(j,k,p,q,s,t)| - \alpha(j,k,p,q,s,t)).$$

Using the conditions of almost regularity (since A is almost strongly regular) and condition (ii), we get

$$L^{\star}(Ax) \leq P\text{-}\limsup_{p,q\to\infty} \sup_{s,t\geq 0} \sum_{j=M+1}^{\infty} \sum_{k=N+1}^{\infty} |\alpha(j,k,p,q,s,t)| x_{jk}. \qquad (5.9)$$

Now

$$\sum_{j=M+1}^{\infty} \sum_{k=N+1}^{\infty} |\alpha(j,k,p,q,s,t)| x_{jk}$$

$$= \sum_{j=M+1}^{\infty} \sum_{k=N+1}^{\infty} (|\alpha(j,k,p,q,s,t)| - \alpha(j-M-1,k-N-1,p,q,s,t)) x_{jk}$$

$$+ \sum_{j=M+1}^{\infty} \sum_{k=N+1}^{\infty} (\alpha(j-M-1,k-N-1,p,q,s,t)) x_{jk}$$

$$\leq \Sigma' + (L^{\star}(x)+\epsilon) \sum_{j=0}^{\infty} \sum_{k=0}^{\infty} \alpha(j,k,p,q,s,t), \quad \text{by (5.8),} \qquad (5.10)$$

where

$$\Sigma' = \sum_{j=M+1}^{\infty} \sum_{k=N+1}^{\infty} (|\alpha(j,k,p,q,s,t)| - \alpha(j-M-1,k-N-1,p,q,s,t)) x_{jk}$$

$$\leq \|x\| \sum_{i=1}^{M+1} \sum_{r=1}^{N+1} \left((M-i+1) \sum_{j=0}^{\infty} \sum_{k=0}^{\infty} |\Delta_{10}\alpha(j,k,p,q,s,t)| \right.$$

$$\left. + (N-r+1) \sum_{j=0}^{\infty} \sum_{k=0}^{\infty} |\Delta_{01}\alpha(j,k,p,q,s,t)| \right). \qquad (5.11)$$

By (5.9), (5.10), and (5.11), using the conditions of almost strong regularity, we get $L^{\star}(Ax) \leq L^{\star}(x)$ since ϵ is arbitrary. $\square$

5.4 Examples

5.4.1 Almost Convergent Sequences

(i) Define the double sequence $x = (x_{jk})$ by

$$x_{jk} = \begin{cases} 1 & \text{if } j \text{ is odd, for all } k, \\ 0 & \text{otherwise.} \end{cases}$$

Then x is almost convergent to $\frac{1}{2}$.

(ii) Define $x = (x_{jk})$ by

$$x_{jk} = (-1)^j \quad \text{for all } k.$$

Then x is almost convergent to 0.

5.4.2 Strongly Regular Matrix

Define $A = (a_{jk})$ by

$$a_{mnjk} = \begin{cases} \frac{1}{m^2} & \text{if } m = n \text{ and } j, k \le m \text{ (even),} \\ \frac{1}{m^2 - m} & \text{if } m = n, \ j \ne k, \text{ and } j, k \le m \text{ (odd),} \\ 0 & \text{otherwise.} \end{cases}$$

We can easily verify that A is strongly regular, that is, conditions (RH_1)–(RH_5), (3.4), and (3.5) of Chap. 3 hold. Moreover, for the sequence defined as in Sect. 5.4.1(i), we have

$$\sum_{j=1}^{\infty}\sum_{k=1}^{\infty} a_{mnjk} x_{jk} = a_{mm11}x_{11} + a_{mm12}x_{12} + \cdots + a_{mm1m}x_{1m}$$

$$+ a_{mm21}x_{21} + a_{mm22}x_{22} + \cdots + a_{mm2m}x_{2m}$$

$$+ a_{mm31}x_{31} + a_{mm32}x_{32} + a_{mm33}x_{33} + \cdots + a_{mm3m}x_{3m}$$

$$\vdots$$

$$+ a_{mmm-1,1}x_{m-1,1} + \cdots + a_{mmm-1,m}x_{m-1,m}$$

$$+ a_{mmm1}x_{m1} + \cdots + a_{mmmm}x_{mm}$$

$$= \frac{m}{m^2} \cdot \frac{m}{2}, \quad \text{if } m \text{ is even,}$$

$$\rightarrow \frac{1}{2} \quad \text{as } m, n \rightarrow \infty.$$

Similarly,

$$\sum_{j=1}^{\infty}\sum_{k=1}^{\infty} a_{mnjk} x_{jk} = \frac{m-1}{m^2-m} \cdot \frac{m+1}{2} \quad \text{if } m \text{ is odd,}$$

$$\rightarrow \frac{1}{2} \quad \text{as } m, n \rightarrow \infty,$$

that is,

$$BP\text{-}\lim Ax = \frac{1}{2} = \mathcal{F}\text{-}\lim x,$$

and so A transforms almost convergent sequence into convergent (BP-convergent) to the same limit.

5.4.3 Bounded-Regular Matrix that Is Not Strongly Regular

In Sect. 5.4.2, A is strongly regular and so bounded-regular. Let us define a four-dimensional matrix $A = (a_{mnjk})$ as

$$a_{mnjk} = \begin{cases} \frac{2}{m^2} & \text{if } m = n, \ j+k = \text{even and } j, k \leq m \text{ (even)}, \\ \frac{1}{m^2-m} & \text{if } m = n, j \neq k \text{ and } j, k \leq m \text{ (odd)}, \\ 0 & \text{otherwise.} \end{cases}$$

Then A is bounded-regular but not strongly regular. Conditions (RH_1)–(RH_5) can easily be verified. But

$$\lim_{m,n} \sum_{j=1}^{\infty}\sum_{k=1}^{\infty} |a_{mnjk} - a_{m,n,j+1,k}| = \begin{cases} 2 & \text{if } m \text{ is even,} \\ 0 & \text{if } m \text{ is odd,} \end{cases}$$

and also

$$\lim_{m,n} \sum_{j=1}^{\infty}\sum_{k=1}^{\infty} |a_{mnjk} - a_{m,n,j,k+1}| = \begin{cases} 2 & \text{if } m \text{ is even,} \\ 0 & \text{if } m \text{ is odd.} \end{cases}$$

Therefore, conditions (3.4) and (3.5) of Chap. 3 do not hold, and so A is not strongly regular.

5.4.4 In Theorem 5.6, Strong Regularity of A Cannot Be Replaced by Bounded-Regularity

Consider the matrix $A = (a_{mnjk})$ as defined in Sect. 5.4.3. This is bounded-regular but not strongly regular, and also

$$P\text{-}\lim_{m,n} \sum_{j,k=0,0}^{\infty,\infty} |a_{mnjk}| = 1,$$

i.e., condition (ii) of Theorem 5.6 holds. Take the bounded double sequence $x = (x_{jk})$ defined by $x_{jk} = (-1)^{j+k}$, which is almost convergent to zero, that is,

$$L^{\star}(x) = 0.$$

Now

$$\sum_{j,k} a_{mnjk} x_{jk} = \begin{cases} \frac{2}{m^2} \cdot \frac{m}{2} \cdot m & \text{if } m \text{ is even,} \\ \frac{-1}{m^2-m} \cdot m & \text{if } m \text{ is odd.} \end{cases}$$

Therefore,

$$P\text{-}\limsup_{m,n} \sum_{j,k} a_{mnjk} x_{jk} = 1$$

and

$$P\text{-}\liminf_{m,n} \sum_{j,k} a_{mnjk} x_{jk} = 0,$$

i.e., $L(Ax) = 1$. Hence, $L(Ax) > L^{\star}(x)$, that is, (5.4) does not hold.

5.5 Riesz Core

In this section, we define the Riesz core by using the concept of Riesz convergence [3].

Definition 5.8 The Riesz core (or R-core) of a double sequence $x = (x_{jk})$ is the closed interval $[P\text{-}\liminf_{m,n} t_{mn}^{qp}(x), P\text{-}\limsup_{m,n} t_{mn}^{qp}(x)]$.

Note that in the case $q_i = 1$ for all i and $p_j = 1$ for all j, Riesz core is reduced to the Cesàro core [53].

Now, we establish the following inequality.

Theorem 5.9 Let $\|A\| < \infty$. Then,

$$R\text{-}core\{Ax\} \subseteq P\text{-}core\{x\},$$

i.e.,

$$P\text{-}\limsup t_{rs}^{qp}(Ax) \le P\text{-}\limsup x, \qquad (5.12)$$

for all $x \in \mathcal{M}_u$ if and only if $A \in (\mathcal{C}_{BP}, \mathcal{R}_{BP})_{\text{reg}}$ and

$$P\text{-}\lim_{m,n} \sum_{r,s} |\eta(m,n,r,s,q,p)| = 1, \qquad (5.13)$$

where $\eta(m, n, r, s, q, p)$ is the same as in Theorem 4.9, i.e.,

$$\eta(m, n, r, s, q, p) = \frac{1}{Q_r} \frac{1}{P_s} \sum_{j=1}^{r} \sum_{k=1}^{s} q_j p_k a_{mnjk}.$$

Proof Let (5.12) hold for all $x \in \mathcal{M}_u$. Then, it is easy to get that

$$-P\text{-}\limsup(-x) \leq -P\text{-}\limsup t_{rs}^{qp}(-Ax) \leq P\text{-}\limsup t_{rs}^{qp}(Ax) \leq P\text{-}\limsup x.$$

Since

$$-P\text{-}\limsup(-x) = P\text{-}\liminf x \quad \text{and} \quad -P\text{-}\limsup t_{rs}^{qp}(-Ax) = P\text{-}\liminf t_{rs}^{qp}(Ax),$$

by choosing $x \in \mathcal{C}_{BP}$, we reach that $P\text{-}\lim t_{rs}^{qp}(Ax) = P\text{-}\lim(x)$. Since x is arbitrary, this means that $A \in (\mathcal{C}_{BP}, \mathcal{R}_{BP})_{\text{reg}}$.

By Theorem 4.3, there exists $y \in \mathcal{M}_u$ with $\|y\| \leq 1$ such that

$$P\text{-}\limsup t_{rs}^{qp}(Ay) = P\text{-}\limsup \sum_{r,s} |\eta(m, n, r, s, q, p)|.$$

So, from the assumption we have that

$$P\text{-}\limsup \sum_{r,s} |\eta(m, n, r, s, q, p)| = P\text{-}\limsup t_{rs}^{qp}(Ay) \leq P\text{-}\limsup(y) \leq \|y\| \leq 1.$$

By the same way, since one can see that

$$P\text{-}\liminf \sum_{r,s} |\eta(m, n, r, s, q, p)| \geq 1,$$

we get the necessity of (5.13).

Conversely, suppose that $A \in (\mathcal{C}_{BP}, \mathcal{R}_{BP})_{\text{reg}}$ and (5.13) holds. For any arbitrary bounded sequence $x = (x_{rs})$, there exist $M, N > 0$ such that $x_{rs} \leq P\text{-}\limsup x + \varepsilon$ whenever $r > M, s > N$. Now, we can write the following inequality:

$$\left| \sum_{r=0}^{\infty} \sum_{s=0}^{\infty} \eta(m, n, r, s, q, p) x_{rs} \right|$$

$$= \left| \sum_{r=0}^{\infty} \sum_{s=0}^{\infty} \left(\frac{|\eta(m, n, r, s, q, p)| + \eta(m, n, r, s, q, p)}{2} \right. \right.$$

$$\left. \left. - \frac{|\eta(m, n, r, s, q, p)| - \eta(m, n, r, s, q, p)}{2} \right) x_{rs} \right|$$

$$\leq \sum_{r=0}^{\infty} \sum_{s=0}^{\infty} |\eta(m, n, r, s, q, p)| |x_{rs}|$$

$$+ \sum_{r=0}^{\infty} \sum_{s=0}^{\infty} \left| \left(\left| \eta(m,n,r,s,q,p) \right| - \eta(m,n,r,s,q,p) \right) x_{rs} \right|$$

$$\leq \|x\| \sum_{r=0}^{M} \sum_{s=0}^{N} \left| \eta(m,n,r,s,q,p) \right| + \|x\| \sum_{r=M+1}^{\infty} \sum_{s=0}^{N} \left| \eta(m,n,r,s,q,p) \right|$$

$$+ \|x\| \sum_{r=0}^{M} \sum_{s=N+1}^{\infty} \left| \eta(m,n,r,s,q,p) \right|$$

$$+ (P\text{-}\lim\sup x + \varepsilon) \sum_{r=M+1}^{\infty} \sum_{s=N+1}^{\infty} \left| \eta(m,n,r,s,q,p) \right|$$

$$+ \|x\| \sum_{r=0}^{\infty} \sum_{s=0}^{\infty} \left(\left| \eta(m,n,r,s,q,p) \right| - \eta(m,n,r,s,q,p) \right).$$

Using the conditions of the class $(\mathcal{C}_{BP}, \mathcal{R}_{BP})_{\mathrm{reg}}$ and (5.13), we reach that $P\text{-}\lim\sup t_{rs}^{qp}(Ax) \leq P\text{-}\lim\sup x$. □

5.6 Exercises

1 Determine the conditions for

$$P\text{-core}\{Ax\} \subseteq R\text{-core}\{x\}.$$

2 Find the necessary and sufficient conditions for

$$R\text{-core}\{Ax\} \subseteq M\text{-core}\{x\}.$$

3 Obtain the conditions for

$$M\text{-core}\{Ax\} \subseteq R\text{-core}\{x\}.$$

4 Find the necessary and sufficient conditions for

$$R\text{-core}\{Ax\} \subseteq R\text{-core}\{x\}.$$

5 For an arbitrary matrix $A = (a_{mnjk})$, in order that, whenever $Bx \in \mathcal{M}_u$, Ax should exist and be bounded. Find necessary and sufficient conditions for

$$P\text{-core}\{Ax\} \subseteq P\text{-core}\{Bx\},$$

where $B = (b_{mnjk})$ is a normal matrix (i.e., triangular with nonzero diagonal entries), and denote its triangular inverse by $B^{-1} = (b_{mnjk}^{-1})$.

6 Determine the conditions for

$$R\text{-core}\{Ax\} \subseteq P\text{-core}\{Bx\}.$$

7 Determine the conditions for

$$R\text{-core}\{Ax\} \subseteq M\text{-core}\{Bx\}.$$

Chapter 6
Application of Almost Convergence in Approximation Theorems for Functions of Two Variables

In this chapter, we apply the notion of almost convergence for double sequences to prove some Korovkin-type approximation theorems for functions of two variables through some different sets of test functions. We also give examples in support of our results, and furthermore we present some consequences of the main results.

6.1 Introduction

Korovkin-type approximation theorems are useful tools to check whether a given sequence $(L_n)_{n \geq 1}$ of positive linear operators on the space $C[0, 1]$ of all continuous functions on the real interval $[0, 1]$ is an approximation process. That is, these theorems exhibit a variety of test functions that assure that the approximation property holds on the whole space if it holds for them. Such a property was discovered by Korovkin in 1953 for the functions 1, x, and x^2 in the space $C[0, 1]$ and for the functions 1, $\cos x$, and $\sin x$ in the space of all continuous 2π-periodic functions on the real line.

Let $C[a, b]$ be the space of all functions f continuous on $[a, b]$. We know that $C[a, b]$ is a Banach space with norm

$$\|f\|_\infty := \sup_{x \in [a,b]} |f(x)|, \quad f \in C[a, b].$$

The classical Korovkin approximation theorem states as follows.

Let (T_n) be a sequence of positive linear operators from $C[a, b]$ into $C[a, b]$. Then $\lim_n \|T_n(f, x) - f(x)\|_\infty = 0$ for all $f \in C[a, b]$ if and only if $\lim_n \|T_n(f_i, x) - f_i(x)\|_\infty = 0$ for $i = 0, 1, 2$, where $f_0(x) = 1$, $f_1(x) = x$, and $f_2(x) = x^2$.

Mohapatra [72] was the first to use the notion of almost convergence for ordinary sequences to prove some approximation results. Quite recently, approximation theorems of such type are proved in [6, 7], and [73] for almost convergence of double and single sequences, respectively. In this chapter, we use the notion of almost convergence of double sequences to prove Korovkin-type approximation theorems for functions of two variables through different sets of test functions.

M. Mursaleen, S.A. Mohiuddine, *Convergence Methods for Double Sequences and Applications*, DOI 10.1007/978-81-322-1611-7_6, © Springer India 2014

6.2 For Test Functions $1, x, y, x^2 + y^2$

Let $C(I^2)$ be the space of all two-dimensional continuous real-valued functions on $I \times I$, where $I = [a, b]$. Suppose that $T_{m,n} : C(I^2) \to C(I^2)$. We write $T_{m,n}(f; x, y)$ for $T_{m,n}(f(s, t); x, y)$, and we say that T is a positive operator if $T(f; x, y) \geq 0$ for all $f(x, y) \geq 0$.

The following version of the classical Korovkin approximation theorem was given by Volkov [124].

Theorem 6.1 *Let $(T_{j,k})$ be a double sequence of positive linear operators from $C(I^2)$ into $C(I^2)$. Then, for all $f \in C(I^2)$,*

$$\lim_{j,k \to \infty} \left\| T_{j,k}(f; x, y) - f(x, y) \right\|_\infty = 0$$

if and only if

$$\lim_{j,k \to \infty} \left\| T_{j,k}(f_i; x, y) - f_i(x, y) \right\|_\infty = 0 \quad (i = 0, 1, 2, 3),$$

where $f_0(x, y) = 1$, $f_1(x, y) = x$, $f_2(x, y) = y$, and $f_3(x, y) = x^2 + y^2$.

We prove the following theorem for almost convergence.

Theorem 6.2 *Let $(T_{j,k})$ be a double sequence of positive linear operators from $C(I^2)$ into $C(I^2)$ and $D_{m,n,p,q}(f; x, y) = \frac{1}{pq} \sum_{j=m}^{m+p-1} \sum_{k=n}^{n+q-1} T_{j,k}(f; x, y)$. Then, for all $f \in C(I^2)$,*

$$\mathcal{F}\text{-} \lim_{j,k \to \infty} \left\| T_{j,k}(f; x, y) - f(x, y) \right\|_\infty = 0, \quad i.e.,$$

$$\lim_{p,q \to \infty} \left\| D_{m,n,p,q}(f; x, y) - f(x, y) \right\|_\infty = 0 \quad \text{uniformly in } m, n, \tag{6.1}$$

if and only if

$$\lim_{p,q \to \infty} \left\| D_{m,n,p,q}(1; x, y) - 1 \right\|_\infty = 0 \quad \text{uniformly in } m, n, \tag{6.2}$$

$$\lim_{p,q \to \infty} \left\| D_{m,n,p,q}(s; x, y) - x \right\|_\infty = 0 \quad \text{uniformly in } m, n, \tag{6.3}$$

$$\lim_{p,q \to \infty} \left\| D_{m,n,p,q}(t; x, y) - y \right\|_\infty = 0 \quad \text{uniformly in } m, n, \tag{6.4}$$

$$\lim_{p,q \to \infty} \left\| D_{m,n,p,q}(s^2 + t^2; x, y) - (x^2 + y^2) \right\|_\infty = 0 \quad \text{uniformly in } m, n. \tag{6.5}$$

Proof Since each of the functions $1, x, y, x^2 + y^2$ belongs to $C(I^2)$, conditions (6.2)–(6.5) follow immediately from (6.1). By the continuity of f on I^2, we can write $|f(x, y)| \leq M, a \leq x, y \leq b$, where $M = \|f\|_\infty$. Therefore,

$$\left| f(s, t) - f(x, y) \right| \leq 2M, \quad a \leq s, t, x, y \leq b. \tag{6.6}$$

Also, since $f \in C(I^2)$, for every $\epsilon > 0$, there is $\delta > 0$ such that

$$|f(s,t) - f(x,y)| < \epsilon \quad \forall |s - x| < \delta \text{ and } |t - y| < \delta. \tag{6.7}$$

Using (6.6), (6.7) and putting $\psi_1 = \psi_1(s, x) = (s - x)^2$ and $\psi_2 = \psi_2(t, y) = (t - y)^2$, we get

$$|f(s,t) - f(x,y)| < \epsilon + \frac{2M}{\delta^2}(\psi_1 + \psi_2) \quad \forall |s - x| < \delta \text{ and } |t - y| < \delta,$$

that is,

$$-\epsilon - \frac{2M}{\delta^2}(\psi_1 + \psi_2) < f(s,t) - f(x,y) < \epsilon + \frac{2M}{\delta^2}(\psi_1 + \psi_2).$$

Now, operate $T_{j,k}(1; x, y)$ to this inequality. Since $T_{j,k}(f; x, y)$ is monotone and linear, we obtain

$$T_{j,k}(1; x, y)\left(-\epsilon - \frac{2M}{\delta^2}(\psi_1 + \psi_2)\right) < T_{j,k}(1; x, y)(f(s,t) - f(x,y))$$

$$< T_{j,k}(1; x, y)\left(\epsilon + \frac{2M}{\delta^2}(\psi_1 + \psi_2)\right).$$

Note that x and y are fixed, and so $f(x, y)$ is a constant number. Therefore,

$$-\epsilon T_{j,k}(1; x, y) - \frac{2M}{\delta^2}T_{j,k}(\psi_1 + \psi_2; x, y)$$

$$< T_{j,k}(f; x, y) - f(x, y)T_{j,k}(1; x, y)$$

$$< \epsilon T_{j,k}(1; x, y) + \frac{2M}{\delta^2}T_{j,k}(\psi_1 + \psi_2; x, y). \tag{6.8}$$

But

$$T_{j,k}(f; x, y) - f(x, y)$$
$$= T_{j,k}(f; x, y) - f(x, y)T_{j,k}(1; x, y) + f(x, y)T_{j,k}(1; x, y) - f(x, y)$$
$$= \left[T_{j,k}(f; x, y) - f(x, y)T_{j,k}(1; x, y)\right] + f(x, y)\left[T_{j,k}(1; x, y) - 1\right]. \tag{6.9}$$

Using (6.8) and (6.9), we have

$$T_{j,k}(f; x, y) - f(x, y)$$
$$< \epsilon T_{j,k}(1; x, y) + \frac{2M}{\delta^2}T_{j,k}(\psi_1 + \psi_2; x, y) + f(x, y)\left(T_{j,k}(1; x, y) - 1\right). \tag{6.10}$$

Now

$$T_{j,k}(\psi_1 + \psi_2; x, y) = T_{j,k}\left((s - x)^2 + (t - y)^2; x, y\right)$$

$$= T_{j,k}\left(s^2 - 2sx + x^2 + t^2 - 2ty + y^2; x, y\right)$$

$$= T_{j,k}\left(s^2 + t^2; x, y\right) - 2xT_{j,k}(s; x, y) - 2yT_{j,k}(t; x, y)$$

$$+ \left(x^2 + y^2\right)T_{j,k}(1; x, y)$$

$$= \left[T_{j,k}\left(s^2 + t^2; x, y\right) - \left(x^2 + y^2\right)\right] - 2x\left[T_{j,k}(s; x, y) - x\right]$$

$$- 2y\left[T_{j,k}(t; x, y) - y\right] + \left(x^2 + y^2\right)\left[T_{j,k}(1; x, y) - 1\right].$$

Using (6.10), we obtain

$$T_{j,k}(f; x, y) - f(x, y)$$

$$< \epsilon T_{j,k}(1; x, y) + \frac{2M}{\delta^2}\left\{\left[T_{j,k}\left(s^2 + t^2; x, y\right) - \left(x^2 + y^2\right)\right]\right.$$

$$- 2x\left[T_{j,k}(s; x, y) - x\right] - 2y\left[T_{j,k}(t; x, y) - y\right]$$

$$+ \left(x^2 + y^2\right)\left[T_{j,k}(1; x, y) - 1\right]\right\} + f(x, y)\left(T_{j,k}(1; x, y) - 1\right)$$

$$= \epsilon\left[T_{j,k}(1; x, y) - 1\right] + \epsilon + \frac{2M}{\delta^2}\left\{\left[T_{j,k}\left(s^2 + t^2; x, y\right) - \left(x^2 + y^2\right)\right]\right.$$

$$- 2x\left[T_{j,k}(s; x, y) - x\right] - 2y\left[T_{j,k}(t; x, y) - y\right]$$

$$+ \left(x^2 + y^2\right)\left[T_{j,k}(1; x, y) - 1\right]\right\} + f(x, y)\left(T_{j,k}(1; x, y) - 1\right).$$

Since ϵ is arbitrary, we can write

$$T_{j,k}(f; x, y) - f(x, y)$$

$$\leq \epsilon\left[T_{j,k}(1; x, y) - 1\right] + \frac{2M}{\delta^2}\left\{\left[T_{j,k}\left(s^2 + t^2; x, y\right) - \left(x^2 + y^2\right)\right]\right.$$

$$- 2x\left[T_{j,k}(s; x, y) - x\right] - 2y\left[T_{j,k}(t; x, y) - y\right]$$

$$+ \left(x^2 + y^2\right)\left[T_{j,k}(1; x, y) - 1\right]\right\} + f(x, y)\left(T_{j,k}(1; x, y) - 1\right).$$

Similarly,

$$D_{m,n,p,q}(f; x, y) - f(x, y)$$

$$\leq \epsilon\left[D_{m,n,p,q}(1; x, y) - 1\right] + \frac{2M}{\delta^2}\left\{\left[D_{m,n,p,q}\left(s^2 + t^2; x, y\right) - \left(x^2 + y^2\right)\right]\right.$$

$$- 2x\left[D_{m,n,p,q}(s; x, y) - x\right] - 2y\left[D_{m,n,p,q}(t; x, y) - y\right]$$

$$+ \left(x^2 + y^2\right)\left[D_{m,n,p,q}(1; x, y) - 1\right]\right\} + f(x, y)\left(D_{m,n,p,q}(1; x, y) - 1\right),$$

and, therefore,

$$\left\|D_{m,n,p,q}(f; x, y) - f(x, y)\right\|_\infty$$

$$\leq \left(\epsilon + \frac{2M(a^2 + b^2)}{\delta^2} + M \right) \| D_{m,n,p,q}(1; x, y) - 1 \|_\infty$$

$$- \frac{4Ma}{\delta^2} \| D_{m,n,p,q}(s; x, t) - x \|_\infty$$

$$- \frac{4Mb}{\delta^2} \| D_{m,n,p,q}(t; x, y) - y \|_\infty$$

$$+ \frac{2M}{\delta^2} \| D_{m,n,p,q}(s^2 + t^2; x, y) - (x^2 + y^2) \|_\infty.$$

Letting $p, q \to \infty$ and using (6.2)–(6.5), we get

$$\lim_{p,q \to \infty} \| D_{m,n,p,q}(f; x, y) - f(x, y) \|_\infty = 0, \quad \text{uniformly in } m, n. \qquad \square$$

In the following example, we construct a double sequence of positive linear operators that satisfies the conditions of Theorem 6.2 but does not satisfy the conditions of Theorem 6.1.

Example 6.3 Consider the sequence of classical Bernstein polynomials of two variables [119]

$$B_{m,n}(f; x, y)$$

$$:= \sum_{j=0}^{m} \sum_{k=0}^{n} f\left(\frac{j}{m}, \frac{k}{n} \right) \binom{m}{j} \binom{n}{k} x^j (1-x)^{m-j} y^k (1-y)^{n-k}, \quad 0 \leq x, y \leq 1.$$

Let $P_{m,n} : C(I^2) \to C(I^2)$ be defined by

$$P_{m,n}(f; x, y) = (1 + z_{mn}) B_{m,n}(f; x, y),$$

where (z_{mn}) is a double sequence defined by

$$z_{mn} = \begin{cases} 1 & \text{if } m = n \text{ odd,} \\ -1 & \text{if } m = n \text{ even,} \\ 0, & \text{otherwise.} \end{cases} \tag{6.11}$$

It is easy to see that $z = (z_{mn})$ is almost convergent to zero but not P-convergent. Then

$$B_{m,n}(1; x, y) = 1,$$

$$B_{m,n}(s; x, y) = x,$$

$$B_{m,n}(t; x, y) = y,$$

$$B_{m,n}(s^2 + t^2; x, y) = x^2 + y^2 + \frac{x - x^2}{m} + \frac{y - y^2}{n},$$

and the double sequence $(P_{m,n})$ satisfies conditions (6.2)–(6.5). Hence, we have

$$\mathcal{F}\text{-}\lim_{m,n\to\infty}\|P_{m,n}(f;x,y)-f(x,y)\|_\infty=0.$$

On the other hand, we get $P_{m,n}(f;0,0)=(1+z_{mn})f(0,0)$ since $B_{m,n}(f;0,0)=f(0,0)$, and hence

$$\|P_{m,n}(f;x,y)-f(x,y)\|_\infty\geq|P_{m,n}(f;0,0)-f(0,0)|=z_{mn}|f(0,0)|.$$

We see that $(P_{m,n})$ does not satisfy the conditions of Theorem 6.1 since $\lim_{m,n\to\infty}z_{mn}$ does not exist. Hence, Theorem 6.2 is stronger than the classical Theorem 6.1.

6.3 For Test Functions 1, $\frac{x}{1-x}$, $\frac{y}{1-y}$, $(\frac{x}{1-x})^2+(\frac{y}{1-y})^2$

Let $I=[0,A]$, $J=[0,B]$, $A,B\in(0,1)$, and $K=I\times J$. We denote by $C(K)$ the space of all continuous real-valued functions on K. This space is equipped with the norm

$$\|f\|_{C(K)}:=\sup_{(x,y)\in K}|f(x,y)|,\quad f\in C(K).$$

Let $H_\omega(K)$ denote the space of all real-valued functions f on K such that

$$|f(s,t)-f(x,y)|\leq\omega\left(f;\sqrt{\left(\frac{s}{1-s}-\frac{x}{1-x}\right)^2+\left(\frac{t}{1-t}-\frac{y}{1-y}\right)^2}\right),$$

where ω is the modulus of continuity, i.e.,

$$\omega(f;\delta)=\sup_{(s,t),(x,y)\in K}\{|f(s,t)-f(x,y)|:\sqrt{(s-x)^2+(t-y)^2}\leq\delta\}.$$

It is to be noted that any function $f\in H_\omega(K)$ is continuous and bounded on K. The following result was given by Taşdelen and Erençin [122].

Theorem 6.4 Let $(T_{j,k})$ be a double sequence of positive linear operators from $H_\omega(K)$ into $C(K)$. Then, for all $f\in H_\omega(K)$,

$$\lim_{j,k\to\infty}\|T_{j,k}(f;x,y)-f(x,y)\|_{C(K)}=0$$

if and only if

$$\lim_{j,k\to\infty}\|T_{j,k}(f_i;x,y)-f_i\|_{C(K)}=0\quad(i=0,1,2,3),$$

where

$$f_0(x, y) = 1,$$

$$f_1(x, y) = \frac{x}{1-x},$$

$$f_2(x, y) = \frac{y}{1-y},$$

and

$$f_3(x, y) = \left(\frac{x}{1-x}\right)^2 + \left(\frac{y}{1-y}\right)^2.$$

We prove the following result.

Theorem 6.5 *Let* $(T_{j,k})$ *be a double sequence of positive linear operators from* $H_\omega(K)$ *into* $C(K)$. *Then, for all* $f \in H_\omega(K)$,

$$\mathcal{F}\text{-}\lim \|T_{j,k}(f; x, y) - f(x, y)\|_{C(K)} = 0 \tag{6.12}$$

if and only if

$$\mathcal{F}\text{-}\lim \|T_{j,k}(1; x, y) - 1\|_{C(K)} = 0, \tag{6.13}$$

$$\mathcal{F}\text{-}\lim \left\|T_{j,k}\left(\frac{s}{1-s}; x, y\right) - \frac{x}{1-x}\right\|_{C(K)} = 0, \tag{6.14}$$

$$\mathcal{F}\text{-}\lim \left\|T_{j,k}\left(\frac{t}{1-t}; x, y\right) - \frac{y}{1-y}\right\|_{C(K)} = 0, \tag{6.15}$$

$$\mathcal{F}\text{-}\lim \left\|T_{j,k}\left(\left(\frac{s}{1-s}\right)^2 + \left(\frac{t}{1-t}\right)^2; x, y\right) - \left(\left(\frac{x}{1-x}\right)^2 + \left(\frac{y}{1-y}\right)^2\right)\right\|_{C(K)} = 0. \tag{6.16}$$

Proof Since each of the functions $1, \frac{x}{1-x}, \frac{y}{1-y}, (\frac{x}{1-x})^2 + (\frac{y}{1-y})^2$ belongs to $H_\omega(K)$, conditions (6.13)–(6.16) follow immediately from (6.12). Let $f \in H_\omega(K)$ and $(x, y) \in K$ be fixed. Then, after using the properties of f, a simple calculation gives that

$$|T_{j,k}(f; x, y) - f(x, y)|$$

$$\leq \varepsilon + M\{|T_{j,k}(f_0; x, y) - f_0(x, y)| + |T_{j,k}(f_1; x, y) - f_1(x, y)|$$

$$+ |T_{j,k}(f_2; x, y) - f_2(x, y)| + |T_{j,k}(f_3; x, y) - f_3(x, y)|\},$$

where $N = \|f\|_{C(K)}$, and

$$M = \max\left\{\varepsilon + N + \frac{2N}{\delta^2}\left(\left(\frac{A}{1-A}\right)^2 + \left(\frac{B}{1-B}\right)^2\right),\right.$$

$$\left.\frac{4N}{\delta^2}\left(\frac{A}{1-A}\right), \frac{4N}{\delta^2}\left(\frac{B}{1-B}\right), \frac{2N}{\delta^2}\right\}.$$

Now replacing $T_{j,k}(f; x, y)$ by $\frac{1}{pq}\sum_{j=m}^{m+p-1}\sum_{k=n}^{n+q-1} T_{j,k}(f; x, y) = S_{pq}^{mn}(f; x, y)$ and taking $\sup_{(x,y)\in K}$, we get

$$\left\|S_{pq}^{mn}(f; x, y) - f(x, y)\right\|_{C(K)}$$

$$\leq \varepsilon + M\left(\left\|S_{pq}^{mn}(f_0; x, t) - f_0(x, y)\right\|_{C(K)}\right.$$

$$+ \left\|S_{pq}^{mn}(f_1; x, y) - f_1(x, y)\right\|_{C(K)} + \left\|S_{pq}^{mn}(f_2; x, y) - f_2(x, y)\right\|_{C(K)}$$

$$+ \left.\left\|S_{pq}^{mn}T_{j,k}(f_3; x, y) - f_3(x, y)\right\|_{C(K)}\right). \tag{6.17}$$

Now taking the $\lim_{p,q\to\infty}$ uniformly in m, n on both sides and using conditions (6.13)–(6.16), we get

$$\lim_{p,q\to\infty}\left\|S_{pq}^{mn}(f; x, y) - f(x, y)\right\|_{C(K)} = 0 \quad \text{uniformly in } m, n.$$

Thus, conditions (6.13)–(6.16) imply condition (6.12).   □

We show that the following double sequence of positive linear operators satisfies the conditions of Theorem 6.5 but does not satisfy the conditions of Theorem 6.4.

Example 6.6 Consider the following two-dimensional version of Meyer–König and Zeller operators [70]:

$$B_{m,n}(f; x, y) := (1-x)^{m+1}(1-y)^{n+1}$$

$$\times \sum_{j=0}^{\infty}\sum_{k=0}^{\infty} f\left(\frac{j}{j+m+1}, \frac{k}{k+n+1}\right)\binom{m+j}{j}\binom{n+k}{k}x^j y^k,$$

$$\tag{6.18}$$

where $f \in H_\omega(K)$, and $K = [0, A] \times [0, B]$, $A, B \in (0, 1)$.
   Since, for $x \in [0, A]$, $A \in (0, 1)$,

$$\frac{1}{(1-x)^{m+1}} = \sum_{j=0}^{\infty}\binom{m+j}{j}x^j,$$

it is easy to see that

$$B_{m,n}(f_0; x, y) = f_0(x, y).$$

Also, we obtain

$$B_{m,n}(f_1; x, y) = (1-x)^{m+1}(1-y)^{n+1} \sum_{j=0}^{\infty} \sum_{k=0}^{\infty} \frac{j}{m+1} \binom{m+j}{j} \binom{n+k}{k} x^j y^k$$

$$= (1-x)^{m+1}(1-y)^{n+1} x \sum_{j=0}^{\infty} \sum_{k=0}^{\infty} \frac{1}{m+1} \frac{(m+j)!}{m!(j-1)!} \binom{n+k}{k} x^{j-1} y^k$$

$$= (1-x)^{m+1}(1-y)^{n+1} x \frac{1}{(1-x)^{m+2}} \frac{1}{(1-y)^{n+1}} = \frac{x}{(1-x)},$$

and, similarly,

$$B_{m,n}(f_2; x, y) = \frac{y}{(1-y)}.$$

Finally, we get

$$B_{m,n}(f_3; x, y)$$

$$= (1-x)^{m+1}(1-y)^{n+1}$$

$$\times \sum_{j=0}^{\infty} \sum_{k=0}^{\infty} \left\{ \left(\frac{j}{m+1}\right)^2 + \left(\frac{k}{n+1}\right)^2 \right\} \binom{m+j}{j} \binom{n+k}{k} x^j y^k$$

$$= (1-x)^{m+1}(1-y)^{n+1} \frac{x}{m+1} \sum_{j=0}^{\infty} \sum_{k=0}^{\infty} \frac{j}{m+1} \frac{(m+j)!}{m!(j-1)!} \binom{n+k}{k} x^{j-1} y^k$$

$$+ (1-x)^{m+1}(1-y)^{n+1} \frac{y}{n+1} \sum_{j=0}^{\infty} \sum_{k=0}^{\infty} \frac{k}{n+1} \binom{m+j}{j} \frac{(n+k)!}{n!(k-1)!} x^j y^{k-1}$$

$$= (1-x)^{m+1}(1-y)^{n+1} \frac{x}{m+1} \left\{ x \sum_{j=0}^{\infty} \sum_{k=0}^{\infty} \frac{(m+j+1)!}{(m+1)!(j-1)!} \binom{n+k}{k} x^{j-1} y^k \right.$$

$$+ \left. \sum_{j=0}^{\infty} \sum_{k=0}^{\infty} \binom{m+j+1}{j} \binom{n+k}{k} x^j y^k \right\}$$

$$+ (1-x)^{m+1}(1-y)^{n+1} \frac{y}{n+1} \left\{ y \sum_{j=0}^{\infty} \sum_{k=0}^{\infty} \frac{(n+k+1)!}{(n+1)!(k-1)!} \binom{m+j}{j} x^j y^{k-1} \right.$$

$$+ \left. \sum_{j=0}^{\infty} \sum_{k=0}^{\infty} \binom{n+k+1}{k} \binom{m+j}{j} x^j y^k \right\}$$

$$= \frac{m+2}{m+1} \left(\frac{x}{1-x}\right)^2 + \frac{1}{m+1} \frac{x}{1-x} + \frac{n+2}{n+1} \left(\frac{y}{1-y}\right)^2 + \frac{1}{n+1} \frac{y}{1-y}$$

$$\rightarrow \left(\frac{x}{1-x}\right)^2 + \left(\frac{y}{1-y}\right)^2.$$

Let $L_{m,n} : H_\omega(K) \rightarrow C(K)$ be defined by

$$L_{m,n}(f; x, y) = (1 + z_{mn}) B_{m,n}(f; x, y),$$

where $z = (z_{mn})$ is defined as in (6.11), which is not P-convergent, but $\mathcal{F}$-$\lim z = 0$.

It is easy to see that the sequence $(L_{m,n})$ satisfies conditions (6.13)–(6.16). Hence, by Theorem 6.5 we have

$$\mathcal{F}\text{-} \lim_{m,n\to\infty} \|L_{m,n}(f; x, y) - f(x, y)\| = 0.$$

On the other hand, the sequence $(L_{m,n})$ does not satisfy the conditions of Theorem 6.4 since $(L_{m,n})$ is not P-convergent, that is, Theorem 6.4 does not work for our operators $L_{m,n}$. Hence, our Theorem 6.5 is stronger than Theorem 6.4.

6.4 For Test Functions 1, $\frac{x}{1+x}$, $\frac{y}{1+y}$, $(\frac{x}{1+x})^2 + (\frac{y}{1+y})^2$

Let $I = [0, \infty)$ and $K = I \times I$. We denote by $C_B(K)$ the space of all bounded and continuous real-valued functions on K equipped with the norm

$$\|f\|_{C_B(K)} := \sup_{(x,y)\in K} |f(x, y)|, \quad f \in C_B(K).$$

Let $H_{\omega^*}(K)$ denote the space of all real-valued functions f on K such that

$$|f(s, t) - f(x, y)| \le \omega^*\left(f; \sqrt{\left(\frac{s}{1+s} - \frac{x}{1+x}\right)^2 + \left(\frac{t}{1+t} - \frac{y}{1+y}\right)^2}\right),$$

where ω^* is the modulus of continuity, i.e.,

$$\omega^*(f; \delta) = \sup_{(s,t),(x,y)\in K} \{|f(s, t) - f(x, y)| : \sqrt{(s - x)^2 + (t - y)^2} \le \delta\}.$$

It is to be noted that any function $f \in H_{\omega^*}(K)$ is bounded and continuous on K, and a necessary and sufficient condition for $f \in H_{\omega^*}(K)$ is that

$$\lim_{\delta\to 0} \omega^*(f; \delta) = 0. \tag{6.19}$$

The following is two-dimensional version of the Korovkin-type theorem of Çakar and Gadjiev [21].

Theorem 6.7 *Let (T_{jk}) be a sequence of positive linear operators from $H_{\omega^*}(K)$ into $C_B(K)$. Then, for all $f \in H_{\omega^*}(K)$,*

$$\lim_{j,k \to \infty} \|T_{jk}(f; x, y) - f(x, y)\|_{C_B(K)} = 0 \qquad (6.20)$$

if and only if

$$\lim_{j,k \to \infty} \|T_{jk}(f_i; x, y) - f_i\|_{C_B(K)} = 0 \quad (i = 0, 1, 2, 3), \qquad (6.21)$$

where

$$f_0(x, y) = 1,$$

$$f_1(x, y) = \frac{x}{1+x},$$

$$f_2(x, y) = \frac{y}{1+y},$$

and

$$f_3(x, y) = \left(\frac{x}{1+x}\right)^2 + \left(\frac{y}{1+y}\right)^2.$$

We prove the following result.

Theorem 6.8 *Let (T_{jk}) be a double sequence of positive linear operators from $H_{\omega^*}(K)$ into $C_B(K)$. Then, for all $f \in H_{\omega^*}(K)$,*

$$\mathcal{F}\text{-}\lim \|T_{jk}(f; x, y) - f(x, y)\|_{C_B(K)} = 0 \qquad (6.22)$$

if and only if

$$\mathcal{F}\text{-}\lim \|T_{jk}(1; x, y) - 1\|_{C_B(K)} = 0, \qquad (6.23)$$

$$\mathcal{F}\text{-}\lim \left\|T_{jk}\left(\frac{s}{1+s}; x, y\right) - \frac{x}{1+x}\right\|_{C_B(K)} = 0, \qquad (6.24)$$

$$\mathcal{F}\text{-}\lim \left\|T_{jk}\left(\frac{t}{1+t}; x, y\right) - \frac{y}{1+y}\right\|_{C_B(K)} = 0, \qquad (6.25)$$

$$\mathcal{F}\text{-}\lim \left\|T_{jk}\left(\left(\frac{s}{1+s}\right)^2 + \left(\frac{t}{1+t}\right)^2; x, y\right) - \left(\left(\frac{x}{1+x}\right)^2 + \left(\frac{y}{1+y}\right)^2\right)\right\|_{C_B(K)} = 0. \qquad (6.26)$$

Proof Since each of the functions $f_0(x, y) = 1$, $f_1(x, y) = \frac{x}{1+x}$, $f_2(x, y) = \frac{y}{1+y}$, $f_3(x, y) = (\frac{x}{1+x})^2 + (\frac{y}{1+y})^2$ belongs to $H_{\omega^*}(K)$, conditions (6.23)–(6.26) follow immediately from (6.22). Let $f \in H_{\omega^*}(K)$ and $(x, y) \in K$ be fixed. Then, for $\varepsilon > 0$, there exist $\delta_1, \delta_2 > 0$ such that $|f(s,t) - f(x,y)| < \varepsilon$ for all $(s,t) \in K$ satisfying $|\frac{s}{1+s} - \frac{x}{1+x}| < \delta_1$ and $|\frac{t}{1+t} - \frac{y}{1+y}| < \delta_2$. Let

$$K(\delta) := \left\{ (s,t) \in K : \sqrt{\left(\frac{s}{1+s} - \frac{x}{1+x}\right)^2 + \left(\frac{t}{1+t} - \frac{y}{1+y}\right)^2} < \delta = \min\{\delta_1, \delta_2\} \right\}.$$

Hence,

$$\begin{aligned}
&|f(s,t) - f(x,y)| \\
&= |f(s,t) - f(x,y)|\chi_{K(\delta)}(s,t) + |f(s,t) - f(x,y)|\chi_{K \setminus K(\delta)}(s,t) \\
&\leq \varepsilon + 2N\chi_{K \setminus K(\delta)}(s,t),
\end{aligned} \tag{6.27}$$

where χ_D denotes the characteristic function of the set D, and $N = \|f\|_{C_B(K)}$. Further, we get

$$\chi_{K \setminus K(\delta)}(s,t) \leq \frac{1}{\delta_1^2}\left(\frac{s}{1+s} - \frac{x}{1+x}\right)^2 + \frac{1}{\delta_2^2}\left(\frac{t}{1+t} - \frac{y}{1+y}\right)^2. \tag{6.28}$$

Combining (6.27) and (6.28), we get

$$|f(s,t) - f(x,y)| \leq \varepsilon + \frac{2N}{\delta^2}\left\{\left(\frac{s}{1+s} - \frac{x}{1+x}\right)^2 + \left(\frac{t}{1+t} - \frac{y}{1+y}\right)^2\right\}. \tag{6.29}$$

After using the properties of f, a simple calculation gives that

$$\begin{aligned}
&|T_{jk}(f;x,y) - f(x,y)| \\
&\leq \varepsilon + M\{|T_{jk}(f_0;x,y) - f_0(x,y)| + |T_{jk}(f_1;x,y) - f_1(x,y)| \\
&\quad + |T_{jk}(f_2;x,y) - f_2(x,y)| + |T_{jk}(f_3;x,y) - f_3(x,y)|\},
\end{aligned} \tag{6.30}$$

where

$$M := \varepsilon + N + \frac{4N}{\delta^2}.$$

Now taking $\sup_{(x,y) \in K}$, we get

$$\begin{aligned}
&\|T_{jk}(f;x,y) - f(x,y)\|_{C(K)} \\
&\leq \varepsilon + M\big(\|T_{jk}(f_0;x,t) - f_0(x,y)\|_{C_B(K)} \\
&\quad + \|T_{jk}(f_1;x,y) - f_1(x,y)\|_{C_B(K)} + \|T_{jk}(f_2;x,y) - f_2(x,y)\|_{C_B(K)} \\
&\quad + \|T_{jk}(f_3;x,y) - f_3(x,y)\|_{C_B(K)}\big).
\end{aligned} \tag{6.31}$$

Then, taking $\mathcal{F}$-lim on both sides of (6.31) and using conditions (6.23)–(6.26), we immediately get (6.22). $\qquad \square$

We show that the following double sequence of positive linear operators satisfies the conditions of Theorem 6.8 but does not satisfy the conditions of Theorem 6.7.

Example 6.9 Consider the following operators on two-variable functions (see Bleimann, Butzer, and Hahn [15]):

$$B_{m,n}(f; x, y)$$

$$:= \frac{1}{(1+x)^m (1+y)^n} \sum_{j=0}^{m} \sum_{k=0}^{n} f\left(\frac{j}{m-j+1}, \frac{k}{n-k+1}\right) \binom{m}{j} \binom{n}{k} x^j y^k,$$

where $f \in H_\omega(K)$, $K = [0, \infty) \times [0, \infty)$, and $n \in \mathbb{N}$.
Since

$$(1+x)^m = \sum_{j=0}^{m} \binom{m}{j} x^j \quad \text{and} \quad (1+y)^n = \sum_{k=0}^{n} \binom{n}{k} y^k,$$

it is easy to see that

$$B_{mn}(f_0; x, y) \to 1 = f_0(x, y).$$

Also, by simple calculation we obtain

$$B_{mn}(f_1; x, y) \to \frac{x}{1+x} = f_1(x, y),$$

$$B_{mn}(f_2; x, y) \to \frac{y}{1+y} = f_2(x, y),$$

and

$$B_{mn}(f_3; x, y) \to \left(\frac{x}{1+x}\right)^2 + \left(\frac{y}{1+y}\right)^2 = f_3(x, y).$$

Let the operator $L_{mn} : H_\omega(K) \to C_B(K)$ be defined by

$$L_{mn}(f; x, y) = (1 + z_{mn}) B_{mn}(f; x, y).$$

It is easy to see that the sequence (L_{mn}) satisfies conditions (6.23)–(6.26). Hence, by Theorem 6.8 we have

$$\mathcal{F}\text{-}\lim \|L_{mn}(f; x, y) - f(x, y)\|_{C_B(K)} = 0.$$

On the other hand, the sequence (L_{mn}) does not satisfy the conditions of Theorem 6.7 since (L_{mn}) is almost convergent to 0 but not P-convergent. That is, Theorem 6.7 does not work for our operators L_{mn}. Hence, our Theorem 6.8 is stronger than Theorem 6.7.

6.5 Some Consequences

In this section we present some consequences of Theorems 6.2, 6.5, and 6.8. Note that, throughout this section, we write $t_v(x, y)$ for any set of test functions used in Theorems 6.2, 6.5, or Theorem 6.8.

Theorem 6.10 *Let $(T_{m,n})$ be a double sequence of positive linear operators on $C(I^2)$ such that*

$$\lim_{m,n} \| T_{m+1,n+1} - T_{m,n+1} - T_{m+1,n} + T_{m,n} \| = 0. \tag{6.32}$$

If

$$\mathcal{F}\text{-}\lim_{m,n} \| T_{m,n}(t_v; x, y) - t_v \|_\infty = 0 \quad (v = 0, 1, 2, 3), \tag{6.33}$$

then, for any function $f \in C(I^2)$, we have

$$\lim_{m,n} \| T_{m,n}(f; x, y) - f(x, y) \|_\infty = 0. \tag{6.34}$$

Proof From the above theorems we have that if (6.33) holds, then

$$\lim_{p,q} \| D_{m,n,p,q}(f; x, y) - f(x, y) \|_\infty = 0, \quad \text{uniformly in } m, n. \tag{6.35}$$

We have the following inequality:

$$\| T_{m,n}(f; x, y) - f(x, y) \|_\infty$$

$$\leq \| D_{m,n,p,q}(f; x, y) - f(x, y) \|_\infty$$

$$+ \frac{1}{pq} \sum_{j=m+1}^{m+p-1} \sum_{k=n+1}^{n+q-1} \left(\sum_{\alpha=m+1}^{j} \sum_{\beta=n+1}^{k} \| T_{\alpha,\beta} - T_{\alpha-1,\beta} - T_{\alpha,\beta-1} + T_{\alpha-1,\beta-1} \| \right)$$

$$\leq \| D_{m,n,p,q}(f; x, y) - f(x, y) \|_\infty$$

$$+ \frac{p-1}{2} \frac{q-1}{2} \left\{ \sup_{j \geq m, k \geq n} \| T_{j,k} - T_{j-1,k} - T_{j,k-1} + T_{j-1,k-1} \| \right\}. \tag{6.36}$$

Hence, using (6.32) and (6.35), we get (6.34).   □

We know that double almost convergence implies $(C, 1, 1)$ convergence. This motivates us to further generalize our main result by weakening the hypothesis or to add some condition to get a more general result.

Theorem 6.11 *Let $(T_{m,n})$ be a double sequence of positive linear operators on $C(I^2)$ such that*

$$(C, 1, 1) - \lim_{m,n} \| T_{m,n}(t_v, x) - t_v \|_\infty = 0 \quad (v = 0, 1, 2, 3) \tag{6.37}$$

and

$$\lim_{p,q}\left\{\sup_{m\geq p,n\geq q}\frac{mn}{pq}\left\|\sigma_{m+p-1,n+q-1}(f;x,y)-\sigma_{m-1,n-1}(f;x,y)\right\|_{\infty}\right\}=0,\quad (6.38)$$

where

$$\sigma_{m,n}(f;x,y)=\frac{1}{(m+1)(n+1)}\sum_{j=0}^{m}\sum_{k=0}^{n}T_{j,k}(f;x,y).$$

Then, for any function $f\in C(I^2)$, we have

$$\mathcal{F}\text{-}\lim_{m,n\to\infty}\left\|T_{m,n}(f;x,y)-f(x,y)\right\|_{\infty}=0.$$

Proof For $m\geq p\geq 1$ and $n\geq q\geq 1$, it is easy to show that

$$D_{m,n,p,q}(f;x,y)=\sigma_{m+p-1,n+q-1}(f;x,y)$$
$$+\frac{mn}{pq}\left(\sigma_{m+p-1,n+q-1}(f;x,y)-\sigma_{m-1,n-1}(f;x,y)\right),$$

which implies

$$\sup_{m\geq p,n\geq q}\left\|D_{m,n,p,q}(f;x,y)-\sigma_{m+p-1,n+q-1}(f;x,y)\right\|_{\infty}$$

$$=\sup_{m\geq p,n\geq q}\frac{mn}{pq}\left\|\sigma_{m+p-1,n+q-1}(f;x,y)-\sigma_{m-1,n-1}(f;x,y))\right\|_{\infty}.\quad (6.39)$$

Also by the above theorems, condition (6.37) implies that

$$(C,1,1)\text{-}\lim_{m,n\to\infty}\left\|T_{m,n}(f;x,y)-f(x,y)\right\|_{\infty}=0.\quad (6.40)$$

Using (6.37)–(6.40) and the fact that almost convergence implies $(C,1,1)$ convergence, we get the desired result. $\quad\square$

Theorem 6.12 *Let $(T_{m,n})$ be a double sequence of positive linear operators on $C(I^2)$ such that*

$$\limsup_{m,n}\frac{1}{s,t}\frac{1}{mn}\sum_{j=s}^{s+m-1}\sum_{k=t}^{t+n-1}\|T_{m,n}-T_{j,k}\|=0.$$

If

$$\mathcal{F}\text{-}\lim_{m,n}\left\|T_{m,n}(t_\nu,x)-t_\nu\right\|_{\infty}=0\quad(\nu=0,1,2,3),\quad (6.41)$$

then, for any function $f\in C(I^2)$, we have

$$\lim_{m,n}\left\|T_{m,n}(f;x,y)-f(x,y)\right\|_{\infty}=0.\quad (6.42)$$

Proof From the above theorems we have that if (6.41) holds, then

$$\mathcal{F}\text{-}\lim_{m,n}\left\|T_{m,n}(f;x,y)-f(x,y)\right\|_\infty=0,$$

which is equivalent to

$$\lim_{m,n}\sup_{s,t}\left\|D_{s,t,m,n}(f;x,y)-f(x,y)\right\|_\infty=0. \tag{6.43}$$

Now

$$T_{m,n}-D_{s,t,m,n}=T_{m,n}-\frac{1}{mn}\sum_{j=s}^{s+m-1}\sum_{k=t}^{t+n-1}T_{j,k}$$

$$=\frac{1}{mn}\sum_{j=s}^{s+m-1}\sum_{k=t}^{t+n-1}(T_{m,n}-T_{j,k}).$$

Therefore,

$$\sup_{s,t}\left\|T_{m,n}-D_{s,t,m,n}\right\|_\infty\le\sup_{s,t}\frac{1}{mn}\sum_{j=s}^{s+m-1}\sum_{k=t}^{t+n-1}\left\|T_{m,n}-T_{j,k}\right\|.$$

Now, by using the hypothesis, we get

$$\lim_{m,n}\sup_{s,t}\left\|T_{m,n}(f;x,y)-D_{s,t,m,n}(f;x,y)\right\|_\infty=0. \tag{6.44}$$

By the triangle inequality we have

$$\left\|T_{m,n}(f;x,y)-f(x,y)\right\|_\infty\le\left\|T_{m,n}(f;x,y)-D_{s,t,m,n}(f;x,y)\right\|_\infty$$

$$+\left\|D_{s,t,m,n}(f;x,y)-f(x,y)\right\|_\infty,$$

and, hence, from (6.43) and (6.44) we get

$$\lim_{m,n}\left\|T_{m,n}(f;x,y)-f(x,y)\right\|_\infty=0,$$

that is, (6.42) holds.   □

6.6 Exercises

1 Prove the Korovkin-type approximation theorem for almost convergence of double sequences by using the test functions 1, e^{-x}, e^{-y}, $e^{-2x}+e^{-2y}$.

2 Construct the bivariate polynomial in support of the result in Exercise 1.

3 Prove the Korovkin-type approximation theorem for almost convergence of double sequences by using the test functions 1, $\sin x$, $\sin y$, $\cos x$, $\cos y$.

4 Give an example of a trigonometric polynomial to demonstrate the result of Exercise 3.

5 Prove Theorems 6.10 and 6.11 by considering the test functions 1, e^{-x}, e^{-y}, $e^{-2x} + e^{-2y}$.

6 Prove Theorems 6.10 and 6.11 by considering the test functions 1, $\sin x$, $\sin y$, $\cos x$, $\cos y$.

Chapter 7
Statistical Convergence of Double Sequences

In this chapter, we present the notions of statistical convergence and statistical Cauchy for double sequences $x = (x_{jk})$ introduced and studied by Mursaleen and Edely [93]. We also establish the relation between statistical convergence and strong Cesàro convergence.

7.1 Introduction

First, we define the notion of double natural density to define the concept of statistical convergence of double sequences.

Definition 7.1 Let $K \subseteq \mathbb{N} \times \mathbb{N}$ be a two-dimensional set of positive integers, and let $K(n, m)$ be the numbers of (i, j) in K such that $i \leq n$ and $j \leq m$. Then the two-dimensional analogue of natural density can be defined as follows.

The lower asymptotic density of a set $K \subseteq \mathbb{N} \times \mathbb{N}$ is defined as

$$\delta_2(K) = P\text{-}\liminf_{n,m} \frac{K(n, m)}{nm}.$$

If the sequence $(K(n, m)/nm)$ has a limit in Pringsheim's sense, then we say that K has a double natural density, which is defined as

$$P\text{-}\lim_{n,m} \frac{K(n, m)}{nm} = \delta_2(K).$$

For example, let $K = \{(i^2, j^2) : i, j \in \mathbb{N}\}$. Then

$$\delta_2(K) = P\text{-}\lim_{n,m} \frac{K(n, m)}{nm} \leq P\text{-}\lim_{n,m} \frac{\sqrt{n}\sqrt{m}}{nm} = 0,$$

i.e., the set K has the double natural density zero, while the set $\{(i, 2j) : i, j \in \mathbb{N}\}$ has the double natural density $1/2$.

M. Mursaleen, S.A. Mohiuddine, *Convergence Methods for Double Sequences and Applications*, DOI 10.1007/978-81-322-1611-7_7, © Springer India 2014

Note that, if we set $n = m$, we have a two-dimensional natural density considered by Christopher [23].

7.2 Statistically Convergent Sequences

Mursaleen and Edely [93] (also see [82, 104] and [95]) introduced and studied the statistical analogue for double sequences $x = (x_{jk})$ as follows.

Definition 7.2 A real double sequence $x = (x_{jk})$ is said to be statistically convergent to the number ℓ if for each $\epsilon > 0$, the set

$$\left\{ (i, j), j \leq n \text{ and } k \leq m : |x_{jk} - l| \geq \epsilon \right\}$$

has the double natural density zero. In this case, we write $S\text{-}\lim_{j,k} x_{jk} = \ell$, and we denote the set of all statistically convergent double sequences by S.

Remark 7.3

(a) If x is a convergent double sequence, then it is also statistically convergent to the same number. Since there are only a finite number of bounded (unbounded) rows and/or columns,

$$K(n, m) \leq s_1 n + s_2 m,$$

where s_1 and s_2 are finite numbers, and we can conclude that x is statistically convergent.
(b) If x is statistically convergent to the number l, then l is determined uniquely.
(c) If x is statistically convergent, then x need not be convergent. Also, it is not necessarily bounded. For example, let $x = (x_{jk})$ be defined as

$$x_{jk} = \begin{cases} jk & \text{if } j \text{ and } k \text{ are squares,} \\ 1 & \text{otherwise.} \end{cases} \tag{7.1}$$

It is easy to see that $S\text{-}\lim x_{jk} = 1$ since the cardinality of the set $\{(j, k) : |x_{jk} - 1| \geq \epsilon\} \leq \sqrt{j}\sqrt{k}$ for every $\epsilon > 0$. But x is neither convergent nor bounded.

We prove some analogues for double sequences. For single sequences, such results have been proved by Šalát [112].

Theorem 7.4 *A real double sequence $x = (x_{jk})$ is statistically convergent to a number ℓ if and only if there exists a subset $K = \{(j, k)\} \subseteq \mathbb{N} \times \mathbb{N}$, $j, k = 1, 2, \ldots$, such that $\delta_2(K) = 1$ and*

$$P\text{-}\lim_{\substack{j,k \to \infty \\ (j,k) \in K}} x_{jk} = \ell.$$

Proof Let x be statistically convergent to ℓ. Put

$$K_r = \left\{(j,k) \in \mathbb{N} \times \mathbb{N} : |x_{jk} - \ell| \geq \frac{1}{r}\right\}$$

and

$$M_r = \left\{(j,k) \in \mathbb{N} \times \mathbb{N} : |x_{jk} - \ell| < \frac{1}{r}\right\} \quad (r = 1, 2, 3, \ldots).$$

Then $\delta_2(K_r) = 0$, and

(1) $M_1 \supset M_2 \supset \cdots \supset M_i \supset M_{i+1} \supset \cdots$,

and

(2) $\delta_2(M_r) = 1, r = 1, 2, 3, \ldots$.

Now we have to show that for $(j,k) \in M_r$, (x_{jk}) is convergent to ℓ. Suppose that (x_{jk}) is not convergent to ℓ. Therefore, there is $\epsilon > 0$ such that $|x_{jk} - \ell| \geq \epsilon$ for infinitely many terms. Let

$$M_\epsilon = \left\{(j,k) : |x_{jk} - \ell| < \epsilon\right\} \quad \text{and} \quad \epsilon > \frac{1}{r} \quad (r = 1, 2, \ldots).$$

Then

(3) $\delta_2(M_\epsilon) = 0$,

and, by (1), $M_r \subset M_\epsilon$. Hence, $\delta_2(M_r) = 0$, which contradicts (2). Therefore, (x_{jk}) is convergent to ℓ.

Conversely, suppose that there exists a subset $K = \{(j,k)\} \subseteq \mathbb{N} \times \mathbb{N}$ such that $\delta_2(K) = 1$ and $P\text{-}\lim_{j,k \to \infty, (j,k) \in K} x_{jk} = \ell$, i.e., there exists $N \in \mathbb{N}$ such that for every $\epsilon > 0$,

$$|x_{jk} - \ell| < \epsilon \quad \forall j, k \geq N.$$

Now

$$K_\epsilon = \left\{(j,k) : |x_{jk} - \ell| \geq \epsilon\right\} \subseteq \mathbb{N} \times \mathbb{N} - \left\{(j_{N+1}, k_{N+1}), (j_{N+2}, k_{N+2}), \ldots\right\}.$$

Therefore,

$$\delta_2(K_\epsilon) \leq 1 - 1 = 0.$$

Hence, x is statistically convergent to ℓ. $\qquad\qquad\square$

Remark 7.5 If $S\text{-}\lim_{jk} x_{jk} = \ell$, then there exists a sequence $y = (y_{jk})$ such that $P\text{-}\lim_{j,k} y_{jk} = \ell$ and $\delta_2(\{(j,k) : x_{jk} = y_{jk}\}) = 1$, i.e.,

$$x_{jk} = y_{jk} \quad \text{for almost all } j, k \text{ (for short, a. a. } j, k).$$

Theorem 7.6 *The set $S \cap M_u$ is a closed linear subspace of the normed linear space M_u.*

Proof Let $x^{(nm)} = (x_{jk}^{(nm)}) \in \mathcal{S} \cap \mathcal{M}_u$ and $x^{(nm)} \to x \in \mathcal{M}_u$. Since $x^{(nm)} \in \mathcal{S} \cap \mathcal{M}_u$, there exist real numbers a_{nm} such that

$$\mathcal{S}\text{-}\lim_{j,k} x_{jk}^{(nm)} = a_{nm} \quad (n, m = 1, 2, \ldots).$$

As $x^{(nm)} \to x$, for every $\epsilon > 0$, there exists $N \in \mathbb{N}$ such that

$$\left| x^{(pq)} - x^{(nm)} \right| < \epsilon/3 \tag{7.2}$$

for all $p \geq n \geq N$, $q \geq m \geq N$, where $|\cdot|$ denotes the norm in a linear space.

By Theorem 7.4, there exist subsets K_1 and K_2 of $\mathbb{N} \times \mathbb{N}$ with $\delta_2(K_1) = \delta_2(K_2) = 1$ and

(1) $P\text{-}\lim_{j,k,(j,k) \in K_1} x_{jk}^{(nm)} = a_{nm}$,

(2) $P\text{-}\lim_{j,k,(j,k) \in K_2} x_{jk}^{(pq)} = a_{pq}$.

Now the set $K_1 \cap K_2$ is infinite since $\delta_2(K_1 \cap K_2) = 1$.

Choose $(k_1, k_2) \in K_1 \cap K_2$. We have from (1) and (2) that

$$\left| x_{k_1,k_2}^{(pq)} - a_{pq} \right| < \epsilon/3 \tag{7.3}$$

and

$$\left| x_{k_1,k_2}^{(nm)} - a_{nm} \right| < \epsilon/3. \tag{7.4}$$

Therefore, for all $p \geq n \geq N$ and $q \geq m \geq N$, from (7.2)–(7.4) we have

$$\left| a_{pq} - a_{nm} \right| \leq \left| a_{pq} - x_{k_1,k_2}^{pq} \right| + \left| x_{k_1,k_2}^{pq} - x_{k_1,k_2}^{nm} \right| + \left| x_{k_1,k_2}^{nm} - a_{nm} \right|$$

$$< \frac{\epsilon}{3} + \frac{\epsilon}{3} + \frac{\epsilon}{3} = \epsilon.$$

That is, the sequence (a_{nm}) is a Cauchy sequence and hence convergent. Let

$$P\text{-}\lim_{n,m} a_{nm} = a. \tag{7.5}$$

We need to show that x is statistically convergent to a. Since $x^{(nm)}$ is convergent to x, for every $\epsilon > 0$, there is $N_1(\epsilon)$ such that for $j, k \leq N_1(\epsilon)$,

$$\left| x_{jk}^{(nm)} - x_{jk} \right| < \epsilon/3.$$

Also, from (7.5) we have that for every $\epsilon > 0$, there is $N_2(\epsilon)$ such that for all $j, k \geq N_2(\epsilon)$,

$$|a_{jk} - a| < \epsilon/3.$$

Again, since $x^{(nm)}$ is statistically convergent to a_{nm}, there exists a set $K = \{(j,k)\} \subseteq \mathbb{N} \times \mathbb{N}$ such that $\delta_2(K) = 1$, and for every $\epsilon > 0$, there is $N_3(\epsilon)$ such that for all $j, k \geq N_3(\epsilon)$, $(j, k) \in K$,

$$\left| x_{jk}^{(nm)} - a_{nm} \right| < \epsilon/3.$$

Let $\max\{N_1(\epsilon), N_2(\epsilon), N_3(\epsilon)\} = N_4(\epsilon)$. Then, for a given $\epsilon > 0$ and for all $j, k \geq N_4(\epsilon)$, $(j, k) \in K$,

$$|x_{jk} - a| \leq \left|x_{jk} - x_{jk}^{(nm)}\right| + \left|x_{jk}^{(nm)} - a_{jk}\right| + |a_{jk} - a| < \epsilon/3 + \epsilon/3 + \epsilon/3 = \epsilon.$$

Therefore, x is statistically convergent to a, i.e., $x \in \mathcal{S} \cap \mathcal{M}_u$. Hence, $\mathcal{S} \cap \mathcal{M}_u$ is a closed linear subspace of $\mathcal{M}_u$. $\qquad\square$

Theorem 7.7 *The set $\mathcal{S} \cap \mathcal{M}_u$ is nowhere dense in $\mathcal{M}_u$.*

Proof Since every closed linear subspace of an arbitrary linear normed space S different from S is a nowhere dense set in S [106], we need only to show that $\mathcal{S} \cap \mathcal{M}_u \neq \mathcal{M}_u$. Let us define a double sequence $x = (x_{jk})$ by

$$x_{jk} = \begin{cases} 1 & \text{if } j \text{ and } k \text{ are even,} \\ 0 & \text{otherwise.} \end{cases} \tag{7.6}$$

It is clear that x is not statistically convergent but x is bounded. Hence, $\mathcal{S} \cap \mathcal{M}_u \neq \mathcal{M}_u$. $\qquad\square$

7.3 Statistically Cauchy Sequences

In [45], Fridy has defined the concept of statistically Cauchy single sequences. In this section, we define statistically Cauchy double sequences and prove some analogues.

Definition 7.8 A real double sequence $x = (x_{jk})$ is said to be statistically Cauchy if for every $\epsilon > 0$, there exist $N = N(\epsilon)$ and $M = M(\epsilon)$ such that for all $j, p \geq N$ and $k, q \geq M$, the set

$$\{(j, k), j \leq n, k \leq m : |x_{jk} - x_{pq}| \geq \epsilon\}$$

has the double natural density zero.

Theorem 7.9 *A real double sequence $x = (x_{jk})$ is statistically convergent if and only if x is statistically Cauchy.*

Proof Let x be statistically convergent to a number ℓ. Then for every $\epsilon > 0$, the set

$$\{(j, k), j \leq n, k \leq m : |x_{jk} - \ell| \geq \epsilon\}$$

has the double natural density zero. Choose two numbers N and M such that $|x_{NM} - \ell| \geq \epsilon$. Now let

$$A_\epsilon = \{(j, k), j \leq n, k \leq m : |x_{jk} - x_{NM}| \geq \epsilon\},$$

$$B_\epsilon = \{(j,k), j \le n, k \le m : |x_{jk} - \ell| \ge \epsilon\},$$
$$C_\epsilon = \{(j,k), j = N \le n, k = M \le m : |x_{NM} - \ell| \ge \epsilon\}.$$

Then $A_\epsilon \subseteq B_\epsilon \cup C_\epsilon$, and therefore $\delta_2(A_\epsilon) \le \delta_2(B_\epsilon) + \delta_2(C_\epsilon) = 0$. Hence, x is statistically Cauchy.

Conversely, let x be statistically Cauchy but not statistically convergent. Then there exist N and M such that the set A_ϵ has the double natural density zero. Hence, the set

$$E_\epsilon = \{(j,k), j \le n, k \le m : |x_{jk} - x_{NM}| < \epsilon\}$$

has the double natural density 1. In particular, we can write

$$|x_{jk} - x_{NM}| \le 2|x_{jk} - \ell| < \epsilon \tag{7.7}$$

if $|x_{jk} - \ell| < \epsilon/2$. Since x is not statistically convergent, the set B_ϵ has the double natural density 1, i.e., the set

$$\{(j,k), j \le n, k \le m : |x_{jk} - \ell| < \epsilon\}$$

has the double natural density 0. Therefore, by (7.7) the set

$$\{(j,k), j \le n, k \le m : |x_{jk} - x_{NM}| < \epsilon\}$$

has the double natural density 0, i.e., the set A_ϵ has the natural density 1, which is a contradiction. Hence, x is statistically convergent. $\qquad\square$

We can state the following for double sequences analogous to the result of Fridy [45].

Theorem 7.10 *The following statements are equivalent*:

(a) *x is statistically convergent to ℓ,*
(b) *x is statistically Cauchy,*
(c) *there exists a subsequence y of x such that $P\text{-}\lim_{jk} y_{jk} = \ell$.*

7.4 Relation Between Statistical Convergence and Strong Cesàro Convergence

The following definition of Cesàro summable double sequences is taken from [81].

Definition 7.11 Let $x = (x_{jk})$ be a double sequence. It is said to be Cesàro summable to ℓ if

$$P\text{-}\lim_{n,m} \frac{1}{nm} \sum_{j=1}^{n} \sum_{k=1}^{m} x_{jk} = \ell.$$

We denote the space of all Cesàro summable double sequences by $(C, 1, 1)$.

Similarly, we can define the following as in case of single sequences.

Definition 7.12 Let $x = (x_{jk})$ be a double sequence, and p be a positive real number. Then the double sequence x is said to be strongly p-Cesàro summable to ℓ if

$$P\text{-}\lim_{n,m} \frac{1}{nm} \sum_{j=1}^{n} \sum_{k=1}^{m} |x_{jk} - \ell|^p = 0.$$

We denote the space of all strongly p-Cesàro summable double sequences by w_p^2.

Remark 7.13

(i) If $0 < p \leq q < \infty$, then $w_q^2 \subseteq w_p^2$ (by Hölder's inequality), and

$$w_p^2 \cap \mathcal{M}_u = w_1^2 \cap \mathcal{M}_u \subseteq (C, 1, 1) \cap \mathcal{M}_u.$$

(ii) If x is convergent but unbounded, then x is statistically convergent, but x need be neither Cesàro nor strongly Cesàro summable.

Example 7.14 Let $x = (x_{jk})$ be defined as

$$x_{jk} = \begin{cases} k, & j = 1, \text{ for all } k, \\ j, & k = 1, \text{ for all } j, \\ 0, & \text{otherwise.} \end{cases} \qquad (7.8)$$

Then $P\text{-}\lim_{j,k} x_{jk} = 0$, but

$$P\text{-}\lim_{n,m} \frac{1}{nm} \sum_{j=1}^{n} \sum_{k=1}^{m} x_{jk} = P\text{-}\lim_{n,m} \frac{1}{nm} \frac{1}{2} (m^2 + n^2 + m + n - 2),$$

which does not tend to a finite limit. Hence, x is not Cesàro. Also, x is not strongly Cesàro, but

$$P\text{-}\lim_{n,m} \frac{1}{nm} \left| \{(j, k) : |x_{jk} - 0| \geq \epsilon\} \right| = P\text{-}\lim_{n,m} \frac{m + n - 1}{nm} = 0,$$

i.e., x is statistically convergent to 0.

(iii) If x is a bounded convergent double sequence, then it is also summable by $(C, 1, 1)$, w_p^2, and $\mathcal{S}$.

The following result is an analogue of Theorem 2.1 due to Connor [24].

Theorem 7.15 *(See, e.g., [123].) Let $x = (x_{jk})$ be a double sequence, and p be a positive real number. Then*

(a) *x is statistically convergent to ℓ if it is strongly p-Cesàro summable to ℓ,*
(b) *$w_p^2 \cap \mathcal{M}_u = \mathcal{S} \cap \mathcal{M}_u$.*

Proof (a) Let $K_\epsilon(p) = \{(j,k), j \leq n, k \leq m : |x_{jk} - \ell|^p \geq \epsilon\}$. Now, since x is strongly p-Cesàro summable to ℓ,

$$0 \leftarrow \frac{1}{nm} \sum_{j=1}^{n} \sum_{k=1}^{m} |x_{jk} - \ell|^p$$

$$= \frac{1}{nm} \left\{ \sum_{(j,k) \in K_\epsilon(p)} |x_{jk} - \ell|^p + \sum_{(j,k) \notin K_\epsilon(p)} |x_{jk} - \ell|^p \right\}$$

$$\geq \frac{1}{nm} \left| \{(j,k), j \leq n, k \leq m : |x_{jk} - \ell|^p \geq \epsilon\} \right| \epsilon.$$

Hence, x is statistically convergent to ℓ.

(b) Let

$$I_\epsilon(p) = \left\{ (j,k), j \leq n, k \leq m : |x_{jk} - \ell| \geq (\epsilon/2)^{1/p} \right\}$$

and $M = \|x\|_\infty + |\ell|$, where $\|x\|_\infty$ is the sup-norm for bounded double sequences $x = (x_{jk})$ given in Chap. 2.

Since x is a bounded statistically convergent, we can choose $N = N(\epsilon)$ such that for all $n, m \geq N$,

$$\frac{1}{nm} \left| \left\{ (j,k), j \leq n, k \leq m : |x_{jk} - \ell| \geq \left(\frac{\epsilon}{2}\right)^{1/p} \right\} \right| < \frac{\epsilon}{2M^p}.$$

Now, for all $n, m \geq N$, we have

$$\frac{1}{nm} \sum_{j=1}^{n} \sum_{k=1}^{m} |x_{jk} - \ell|^p = \frac{1}{nm} \left\{ \sum_{(j,k) \in I_\epsilon(p)} |x_{jk} - \ell|^p + \sum_{(j,k) \notin I_\epsilon(p)} |x_{jk} - \ell|^p \right\}$$

$$< \frac{1}{nm} nm \frac{\epsilon}{2M^p} M^p + \frac{1}{nm} nm \frac{\epsilon}{2} = \epsilon.$$

Hence, x is strongly p-Cesàro summable to ℓ. $\qquad\square$

Remark 7.16 Note that if a bounded sequence x is statistically convergent, then it is also $(C, 1, 1)$ summable, but not conversely.

Example 7.17 Let $x = (x_{jk})$ be defined by

$$x_{jk} = (-1)^j \quad \forall k,$$

then

$$P\text{-}\lim_{n,m} \frac{1}{nm} \sum_{j=1}^{n} \sum_{k=1}^{m} x_{jk} = 0,$$

but obviously x is not statistically convergent.

7.5   A-Statistical Convergence and Statistical A-Summability of Double Sequences

Definition 7.18   Let $A = (a_{mnjk})$ be a bounded-regular matrix, and $K \subseteq \mathbb{N} \times \mathbb{N}$. Then A-density of K is defined by $\delta_A^{(2)}(K) := P\text{-}\lim_{m,n} \sum_{(j,k)\in K} a_{mnjk}$, provided that the P-limit exists. A double sequence $x = (x_{jk})$ is said to be A-*statistically convergent* to L if for every $\epsilon > 0$, $\delta_A^{(2)}(K_\epsilon) = 0$, where $K_\epsilon := \{(j,k) \in \mathbb{N} \times \mathbb{N} : |x_{jk} - L| \geq \epsilon\}$. In this case, we write $\mathcal{S}_A\text{-}\lim x = L$.

Note that a P-convergent double sequence is A-statistically convergent to the same value but converse need not be true. Also, note that an A-statistically convergent double sequence need not be bounded.

Definition 7.19   Let $A = (a_{mnjk})$ be a bounded-regular matrix. A double sequence $x = (x_{jk})$ is said to be *statistically A-summable* to L [13] if for every $\epsilon > 0$,

$$\delta_2(\{(m,n) \in \mathbb{N} \times \mathbb{N} : |y_{mn} - L| \geq \epsilon\}) = 0,$$

where

$$y_{mn} = \sum_{j=0}^{\infty} \sum_{k=0}^{\infty} a_{mnjk} x_{jk}.$$

Thus, the double sequence x is statistically A-summable to L if and only if Ax is statistically convergent to L. Now we prove the following relation between A-statistical convergence and statistical A-summability for a double sequence. The case for single sequences has been given in [41].

Theorem 7.20   *If a double sequence $x = (x_{ij})$ is bounded and A-statistically convergent to L, then it is A-summable to L; hence, it is statistically A-summable to L, but not conversely.*

Proof   Let $x = (x_{ij})$ be bounded and A-statistically convergent to L, and $K(\epsilon) = \{(i,j), i \leq m, j \leq n : |x_{ij} - L| \geq \epsilon\}$. Then,

$$|y_{mn} - L| = \left| \sum_{i,j=1,1}^{\infty,\infty} a_{ij}^{mn} x_{ij} - L \right| = \left| \sum_{i,j=1,1}^{\infty,\infty} a_{ij}^{mn}(x_{ij} - L) + L\left(\sum_{i,j=1,1}^{\infty,\infty} a_{ij}^{mn} - 1 \right) \right|$$

$$\leq \left| \sum_{i,j=1,1}^{\infty,\infty} a_{ij}^{mn}(x_{ij} - L) \right| + |L| \left| \sum_{i,j=1,1}^{\infty,\infty} a_{ij}^{mn} - 1 \right|$$

$$\leq \left| \sum_{(i,j)\in K(\epsilon)} a_{ij}^{mn}(x_{ij} - L) \right| + \left| \sum_{(i,j)\notin K(\epsilon)} a_{ij}^{mn}(x_{ij} - L) \right|$$

$$+ |L| \left| \sum_{i,j=1,1}^{\infty,\infty} a_{ij}^{mn} - 1 \right|$$

$$\leq \sup_{i,j} |x_{ij} - L| \sum_{(i,j)\in K(\epsilon)} a_{ij}^{mn} + \epsilon \sum_{i,j\notin K(\epsilon)} a_{ij}^{mn} + |L| \left| \sum_{i,j=1,1}^{\infty,\infty} a_{ij}^{mn} - 1 \right|.$$

Using the definition of A-statistical convergence and the conditions of RH-regularity of A, we get $P\text{-}\lim_{m,n} |y_{mn} - L| = 0$ from the arbitrariness of $\epsilon > 0$. Hence, $S\text{-}\lim_{m,n} |y_{mn} - L| = 0$.

To show that the converse is not true in general, we give the following examples.

Example 7.21

(i) Let $A = (a_{ij}^{mn})$ be four-dimensional Cesàro matrix, i.e.,

$$a_{ij}^{mn} = \begin{cases} 1/mn & \text{if } i \leq m \text{ and } j \leq n, \\ 0 & \text{otherwise}, \end{cases}$$

and let $x = (x_{ij})$ be defined as

$$x_{ij} = (-1)^i \quad \text{for all } j.$$

Then x is $(C, 1, 1)$-summable (and hence statistical $(C, 1, 1)$-summable) to zero but not $(C, 1, 1)$-statistically convergent.

(ii) Define $A = (a_{ij}^{mn})$ by

$$a_{ij}^{mn} = \begin{cases} 1/m^2 & \text{if } m = n, i, j \leq m, \text{ and } m \text{ is even square}, \\ 1/(m^2 - m) & \text{if } m = n, i \neq j, i, j \leq m, \text{ and } m \text{ is odd square}, \\ 0 & \text{otherwise}, \end{cases}$$

and define the double sequence $x = (x_{ij})$ by

$$x_{ij} = \begin{cases} 1 & \text{if } i \text{ is odd and for all } j, \\ 0 & \text{otherwise}. \end{cases}$$

We can easily verify that A is RH-regular, that is, conditions (RH_1)–(RH_6) hold. Moreover, for the sequence defined above, we have

$$\sum_{i,j=1,1}^{\infty,\infty} a_{ij}^{mn} x_{ij} = \begin{cases} 1/2 & \text{if } m \text{ is even square}, \\ (m+1)/2m & \text{if } m \text{ is odd square}, \\ 0 & \text{otherwise}. \end{cases}$$

Then it is clear that x is not A-summable and hence is not A-statistically convergent, but $S\text{-}\lim_{m,n} y_{mn} = 0$, i.e., x is statistically A-summable to zero.   □

7.6 Statistical Convergence in Locally Solid Riesz Spaces

The notion of statistical convergence has been defined and studied in different setups, e.g., in a locally convex space [67], in topological groups [19, 20], in probabilistic normed space [51, 78, 99, 102, 113] in intuitionistic fuzzy normed spaces [52, 61, 77, 97, 98, 105], and in random 2-normed space [74, 90, 92]. Recently, Maio and Kŏcinac [68] studied this notion in topological and uniform spaces, and Albayrak and Pehlivan [2] in locally solid Riesz spaces [75, 80]. In this chapter, we study statistically convergent, statistically bounded, and statistically Cauchy double sequences in locally solid Riesz spaces.

Let X be a real vector space, and $\leq$ be a partial order on this space. Then, X is said to be an *ordered vector space* if it satisfies the following properties:

(i) if $x, y \in X$ and $y \leq x$, then $y + z \leq x + z$ for each $z \in X$,
(ii) if $x, y \in X$ and $y \leq x$, then $\lambda y \leq \lambda x$ for each $\lambda \geq 0$.

If, in addition, X is a lattice with respect to the partial order, then X is said to be a *Riesz space* (or a *vector lattice*) [126].

For an element x of a Riesz space X, the positive part of x is defined by $x^+ = x \vee \theta = \sup\{x, \theta\}$, the negative part of x by $x^- = (-x) \vee \theta$, and the absolute value of x by $|x| = x \vee (-x)$, where θ is the zero element of X.

A subset S of a Riesz space X is said to be *solid* if $y \in S$ and $|x| \leq |y|$ imply $x \in S$.

A *topological vector space* (X, τ) is a vector space X that has a (linear) topology τ such that the algebraic operations of addition and scalar multiplication in X are continuous. The continuity of addition means that the function $f : X \times X \to X$ defined by $f(x, y) = x + y$ is continuous on $X \times X$, and the continuity of scalar multiplication means that the function $f : \mathbb{C} \times X \to X$ defined by $f(\lambda, x) = \lambda x$ is continuous on $\mathbb{C} \times X$.

Every linear topology τ on a vector space X has a base $\mathcal{N}$ for the neighborhoods of θ satisfying the following properties:

(C_1) Each $Y \in \mathcal{N}$ is a *balanced set*, that is, $\lambda x \in Y$ for all $x \in Y$ and every $\lambda \in \mathbb{R}$ with $|\lambda| \leq 1$.
(C_2) Each $Y \in \mathcal{N}$ is an *absorbing set*, that is, for every $x \in X$, there exists $\lambda > 0$ such that $\lambda x \in Y$.
(C_3) For each $Y \in \mathcal{N}$, there exists some $E \in \mathcal{N}$ with $E + E \subseteq Y$.

A linear topology τ on a Riesz space X is said to be *locally solid* [110] if τ has a base at zero consisting of solid sets. A *locally solid Riesz space* (X, τ) is a Riesz space equipped with a locally solid topology τ.

We shall assume throughout that the symbol $\mathcal{N}_{\text{sol}}$ denotes any base at zero consisting of solid sets and satisfying conditions (C_1), (C_2), and (C_3) in a locally solid topology.

We start with the definition of statistical convergence in a locally solid Riesz space (X, τ).

Definition 7.22 Let (X, τ) be a locally solid Riesz space. Then, a double sequence $x = (x_{jk})$ in X is said to be *statistically τ-convergent* to the number $\xi \in X$ if for every τ-neighborhood U of zero,

$$P\text{-}\lim_{m,n\to\infty} \frac{1}{mn} \left|\left\{(j, k), j \leq m \text{ and } k \leq n : x_{jk} - \xi \notin U\right\}\right| = 0.$$

In this case, we write $S(\tau)\text{-}\lim x = \xi$.

Definition 7.23 Let (X, τ) be a locally solid Riesz space. We say that a double sequence $x = (x_{jk})$ in X is *statistically τ-bounded* if for every τ-neighborhood U of zero, there exists $\lambda > 0$ such that the set

$$\left\{(j, k), j \leq m \text{ and } k \leq n : \lambda x_{jk} \notin U\right\}$$

has the double natural density zero.

Definition 7.24 Let (X, τ) be a locally solid Riesz space. A double sequence $x = (x_{jk})$ in X is *statistically τ-Cauchy* if for every τ-neighborhood U of zero, there exist $N, M \in \mathbb{N}$ such that for all $j, p \geq N$ and $k, q \geq M$, the set

$$\left\{(j, k), j \leq m \text{ and } k \leq n : x_{jk} - x_{pq} \notin U\right\}$$

has the double natural density zero.

7.7 (λ, μ)-Statistical Convergence

Recently, the notion of statistical convergence has been generalized in [104], which is a double sequence version of λ-convergence [103]. Note that λ-convergence is a special case of A-statistical convergence introduced by Kolk [58] and further studied in [71] and [54].

Definition 7.25 Let $\lambda = (\lambda_m)$ and $\mu = (\mu_n)$ be two nondecreasing sequences of positive real numbers tending to ∞ such that

$$\lambda_{m+1} \leq \lambda_m + 1, \qquad \lambda_1 = 0$$

and

$$\mu_{n+1} \leq \mu_n + 1, \qquad \mu_1 = 0.$$

Let $K \subseteq \mathbb{N} \times \mathbb{N}$ be a two-dimensional set of positive integers. Then the (λ, μ)-*density* of K is defined as

$$\delta_{\lambda,\mu}(K) = P\text{-}\lim_{m,n} \frac{1}{\lambda_m \mu_n} \left|\left\{m - \lambda_m + 1 \leq j \leq m, n - \mu_n + 1 \leq k \leq n : (j, k) \in K\right\}\right|,$$

provided that the limit on the right-hand side exists.

In the case $\lambda_m = m$, $\mu_n = n$, the (λ, μ)-density reduces to the natural double density. Also, since $(\lambda_m/m) \leq 1$ and $(\mu_n/n) \leq 1$, we have $\delta_2(K) \leq \delta_{\lambda,\mu}(K)$ for every $K \subseteq \mathbb{N} \times \mathbb{N}$.

Definition 7.26 We define the generalized double de la Vallée-Poussin mean by

$$t_{m,n}(x) = \frac{1}{\lambda_m \mu_n} \sum_{j \in J_m} \sum_{k \in I_n} x_{jk},$$

where $J_m = [m - \lambda_m + 1, m]$ and $I_n = [n - \mu_n + 1, n]$.

A double sequence $x = (x_{jk})$ is said to be *strongly* (V, λ, μ)-*summable* to a number ℓ if

$$P\text{-}\lim_{m,n} t_{m,n}\big(|x - \ell \mathbf{e}|\big) = 0.$$

We denote the set of all double strongly (V, λ, μ)-summable sequences by $[V, \lambda, \mu]$. If $\lambda_m = m$ for all m and $\mu_n = n$ for all n, then strongly (V, λ, μ)-summability is reduced to the strongly Cesàro summability, and $[V, \lambda, \mu] = [C, 1, 1]$, the space of strongly Cesàro summable double sequences.

Definition 7.27 A double sequence $x = (x_{jk})$ is said to be (λ, μ)-*statistically convergent* to ℓ if $\delta_{\lambda,\mu}(E) = 0$, where $E = \{j \in J_m, k \in I_n : |x_{jk} - \ell| \geq \epsilon\}$, i.e., if for every $\epsilon > 0$,

$$P\text{-}\lim_{m,n} \frac{1}{\lambda_m \mu_n}\big|\{j \in J_m, k \in I_n : |x_{jk} - \ell| \geq \epsilon\}\big| = 0.$$

In this case, we write $st_{\lambda,\mu}\text{-}\lim_{j,k} x_{jk} = \ell$ and denote the set of all (λ, μ)-statistically convergent double sequences by $S_{\lambda,\mu}$.

Here note that if $\lambda_m = m$ for all m and $\mu_n = n$ for all n, then the space $S_{\lambda,\mu}$ is reduced to the space S.

We write $(S_{\lambda,\mu})_0$ to denote the space of all sequences that are (λ, μ)-statistically convergent to zero and $S_{\lambda,\mu}^\infty$ for bounded (λ, μ)-statistically convergent double sequences, and we write $(S_{\lambda,\mu}^\infty)_0$ for double sequences that are bounded and (λ, μ)-statistically convergent to zero. The following result is a double sequence version of Theorem 2.1 of [103].

Theorem 7.28 *Let λ and μ be the sequences as defined above. Then*

(i) $x_{jk} \to \ell[V, \lambda, \mu]$ *implies* $x_{jk} \to \ell(S_{\lambda,\mu})$, *but not conversely*;
(ii) *if* $x \in \mathcal{M}_u$ *and* $x_{jk} \to \ell(S_{\lambda,\mu})$, *then* $x_{jk} \to \ell[V, \lambda, \mu]$ *and hence* $x_{jk} \to \ell[C, 1, 1]$;
(iii) $S_{\lambda,\mu}^\infty = [V, \lambda, \mu] \cap \mathcal{M}_u$.

Proof (i) Let $\epsilon > 0$ and $x_{jk} \to L[V, \lambda, \mu]$. We have

$$\sum_{j \in J_m, k \in I_n} |x_{jk} - \ell| \geq \sum_{\substack{j \in J_m, k \in I_n \\ |x_{jk} - L| \geq \epsilon}} |x_{jk} - \ell| \geq \epsilon \left| \{ j \in J_m, k \in I_n : |x_{jk} - \ell| \geq \epsilon \} \right|.$$

Hence, $x_{jk} \to \ell(S_{\lambda, \mu})$.

For the converse, let x be defined by

$$x_{jk} = \begin{cases} jk & \text{for } m - [\lambda_m] + 1 \leq j \leq m \text{ and } n - [\mu_n] + 1 \leq k \leq n, \\ 0 & \text{otherwise.} \end{cases}$$

It is clear that x is an unbounded double sequence and for $\epsilon > 0$,

$$P\text{-}\lim_{m,n} \frac{1}{\lambda_m \mu_n} \left| \{ j \in J_m, k \in I_n : |x_{jk} - 0| \geq \epsilon \} \right| = P\text{-}\lim_{m,n} \frac{[\sqrt{\lambda_m \mu_n}]}{\lambda_m \mu_n} = 0.$$

Therefore, $x_{jk} \to 0 \ (S_{\lambda, \mu})$. Also, note that

$$P\text{-}\lim_{m,n} \frac{1}{\lambda_m \mu_n} \sum_{j \in J_m, k \in I_n} |x_{j,k} - 0|$$

does not exist, i.e., $x_{jk} \nrightarrow 0 \ ([V, \lambda, \mu])$.

(ii) Since $x \in \mathcal{M}_u$, $|x_{jk} - \ell| \leq M$ for all j, k. Also, for given $\epsilon > 0$ and m, n large enough, we obtain

$$\frac{1}{\lambda_m \mu_n} \sum_{j \in J_m, k \in I_n} |x_{jk} - \ell| = \frac{1}{\lambda_m \mu_n} \sum_{\substack{j \in J_m, k \in I_n \\ |x_{jk} - \ell| \geq \epsilon}} |x_{jk} - \ell| + \frac{1}{\lambda_m \mu_n} \sum_{\substack{j \in J_m, k \in I_n \\ |x_{jk} - \ell| < \epsilon}} |x_{jk} - \ell|$$

$$\leq \frac{M}{\lambda_m \mu_n} \left| \{ j \in J_m, k \in I_n : |x_{jk} - \ell| \geq \epsilon \} \right| + \epsilon.$$

Therefore, $x \in \mathcal{M}_u$ and $x_{jk} \to \ell(S_{\lambda, \mu})$, which implies $x_{jk} \to \ell[V, \lambda, \mu]$.

Further, we have

$$\frac{1}{mn} \sum_{j=1}^{m} \sum_{k=1}^{n} (x_{jk} - \ell) = \frac{1}{mn} \sum_{j=1}^{m-\lambda_m} \sum_{k=1}^{n-\mu_n} (x_{jk} - \ell) + \frac{1}{mn} \sum_{j \in J_m} \sum_{k \in I_n} (x_{jk} - \ell)$$

$$\leq \frac{1}{\lambda_m \mu_n} \sum_{j=1}^{m-\lambda_m} \sum_{k=1}^{n-\mu_n} |x_{jk} - \ell| + \frac{1}{\lambda_m \mu_n} \sum_{j \in J_m} \sum_{k \in I_n} |x_{jk} - \ell|$$

$$\leq \frac{2}{\lambda_m \mu_n} \sum_{j \in J_m, k \in I_n} |x_{jk} - \ell|.$$

Hence, $x_{jk} \to \ell[C, 1, 1]$ since $x_{jk} \to \ell[V, \lambda, \mu]$.

(iii) $[V, \lambda, \mu] \cap \mathcal{M}_u = S_{\lambda, \mu}^{\infty}$ follows directly from (i) and (ii). □

In [47], the concepts of statistical boundedness, statistical limit superior, and statistical limit inferior for ordinary (single) sequences were introduced, while in [104], these notions were extended for double sequences.

Definition 7.29 Let

$$B_{\lambda,\mu}(x) = \{b \in \mathbb{R} : \delta_{\lambda,\mu}\{(j,k), j \leq m \text{ and } k \leq n : x_{jk} > b\} \neq 0\},$$

$$A_{\lambda,\mu}(x) = \{a \in \mathbb{R} : \delta_{\lambda,\mu}\{(j,k), j \leq m \text{ and } k \leq n : x_{jk} < a\} \neq 0\}.$$

Then

$$st_{\lambda,\mu}\text{-}\limsup x = \begin{cases} \sup B_{\lambda,\mu}(x), & B_{\lambda,\mu}(x) \neq \emptyset, \\ -\infty, & B_{\lambda,\mu}(x) = \emptyset, \end{cases}$$

and

$$st_{\lambda,\mu}\text{-}\liminf x = \begin{cases} \inf A_{\lambda,\mu}(x), & A_{\lambda,\mu}(x) \neq \emptyset, \\ \infty, & A_{\lambda,\mu}(x) = \emptyset. \end{cases}$$

Definition 7.30 A real number sequence $x = (x_{jk})$ is said to be (λ, μ)-*statistical bounded* if there exists a positive number M such that

$$\delta_{\lambda,\mu}\big(\{(j,k), j \leq m \text{ and } k \leq n : |x_{jk}| > M\}\big) = 0.$$

It is clear that if $x = (x_{jk})$ is (λ, μ)-statistically bounded, then it has both $st_{\lambda,\mu}$-$\liminf$ and $st_{\lambda,\mu}$-$\limsup$.

The following theorem is a consequence of the definitions of $st_{\lambda,\mu}$-$\limsup x$ and $st_{\lambda,\mu}$-$\liminf x$.

Theorem 7.31

(a) $st_{\lambda,\mu}$-$\limsup x = \ell$ *if and only if*

    (i) $\delta_{\lambda,\mu}(\{(j,k), j \leq m \text{ and } k \leq n : x_{jk} > \ell - \epsilon\}) \neq 0$;

    (ii) $\delta_{\lambda,\mu}(\{(j,k), j \leq m \text{ and } k \leq n : x_{jk} > \ell + \epsilon\}) = 0$.

(b) $st_{\lambda,\mu}$-$\liminf x = s$ *if and only if*

    (i) $\delta_{\lambda,\mu}(\{(j,k), j \leq m \text{ and } k \leq n : x_{jk} < s + \epsilon\}) \neq 0$;

    (ii) $\delta_{\lambda,\mu}(\{(j,k), j \leq m \text{ and } k \leq n : x_{jk} < s - \epsilon\}) = 0$.

Remark 7.32 It is easy to see that

(i) For any sequence $x = (x_{jk})$, $st_{\lambda,\mu}$-$\liminf x \leq st_{\lambda,\mu}$-$\limsup x$.

(ii) P-$\liminf x \leq st_{\lambda,\mu}$-$\liminf x \leq st_{\lambda,\mu}$-$\limsup x \leq P$-$\limsup x$ for every bounded sequence $x = (x_{jk})$.

7.8 Exercises

1 (Decomposition Theorem) Prove that a double sequence (x_{jk}) is statically convergent to some number ξ if and only if there exist two sequences (u_{jk}) and (v_{jk}) such that

(i) $x_{jk} = u_{jk} + v_{jk}$, $j, k = 0, 1, 2, \ldots$,

(ii) $P\text{-}\lim_{j,k \to \infty} u_{jk} = \xi$, and

(iii) $P\text{-}\lim_{m,n \to \infty} \frac{1}{(m+1)(n+1)} |\{j \leq m \text{ and } k \leq n : v_{jk} \neq 0\}| = 0$.

Moreover, if (x_{jk}) is bounded, then u_{jk} and v_{jk} are also bounded.

Let (X, τ) be a locally solid Riesz space. Then prove the following:

2 Let $x = (x_{jk})$ and $y = (y_{jk})$ be two double sequences in X. Then the following hold:

(i) If $\mathcal{S}(\tau)\text{-}\lim_{j,k} x_{jk} = \xi_1$ and $\mathcal{S}(\tau)\text{-}\lim_{j,k} x_{jk} = \xi_2$, then $\xi_1 = \xi_2$.

(ii) If $\mathcal{S}(\tau)\text{-}\lim_{j,k} x_{jk} = \xi$, then $\mathcal{S}(\tau)\text{-}\lim_{j,k} \alpha x_{jk} = \alpha\xi$, $\alpha \in \mathbb{R}$.

(iii) If $\mathcal{S}(\tau)\text{-}\lim_{j,k} x_{jk} = \xi$ and $\mathcal{S}(\tau)\text{-}\lim_{j,k} y_{jk} = \eta$, then $\mathcal{S}(\tau)\text{-}\lim_{j,k}(x_{jk} + y_{jk}) = \xi + \eta$.

3 If a double sequence $x = (x_{jk})$ is statistically τ-convergent, then it is statistically τ-bounded.

4 If a double sequence $x = (x_{jk})$ is statistically τ-convergent, then it is statistically τ-Cauchy.

5 A double sequence $x = (x_{jk})$ is statistically τ-convergent to a number ξ if it is $\mathcal{S}^*(\tau)$-convergent to ξ in a locally solid Riesz space (X, τ).

6 Prove that a (λ, μ)-statistically bounded sequence $x = (x_{jk})$ is (λ, μ)-statistically convergent if and only if

$$st_{\lambda,\mu}\text{-}\liminf x = st_{\lambda,\mu}\text{-}\limsup x.$$

7 Prove that $\mathcal{S}\text{-}\lim x = \ell$ implies that $\mathcal{S}_{\lambda,\mu}\text{-}\lim x = \ell$ if and only if

$$\liminf_{m} \frac{\lambda_m}{m} > 0 \quad \text{and} \quad \liminf_{n} \frac{\mu_n}{n} > 0.$$

8 Determine a condition such that $\mathcal{S}_{\lambda,\mu}\text{-}\lim x = \ell$ implies $\mathcal{S}\text{-}\lim x = \ell$.

Chapter 8
Statistical Approximation of Positive Linear Operators

In this chapter, we present some Korovkin-type approximation theorems for functions of two variables via statistical convergence, A-statistical convergence, and statistical A-summability. We also study rates of A-statistical convergence of a double sequence of positive linear operators. Through some concrete examples, we show that the results present in this chapter are stronger than the classical results.

8.1 Introduction

Let $F(\mathbb{R})$ denote the linear space of all real-valued functions defined on $\mathbb{R}$. Let $C(\mathbb{R})$ be the space of all functions f continuous on $\mathbb{R}$. We know that $C(\mathbb{R})$ is a normed space with the norm

$$\|f\|_\infty := \sup_{x\in\mathbb{R}}|f(x)|, \quad f \in C(\mathbb{R}).$$

We denote by $C_{2\pi}(\mathbb{R})$ the space of all 2π-periodic functions $f \in C(\mathbb{R})$, which is a normed spaces with

$$\|f\|_{2\pi} = \sup_{t\in\mathbb{R}}|f(t)|.$$

The classical Korovkin first and second theorems are stated as follows [59, 60].

Theorem I *Let (T_n) be a sequence of positive linear operators from $C[0, 1]$ into $F[0, 1]$. Then $\lim_n \|T_n(f, x) - f(x)\|_\infty = 0$ for all $f \in C[0, 1]$ if and only if $\lim_n \|T_n(f_i, x) - e_i(x)\|_\infty = 0$ for $i = 0, 1, 2$, where $e_0(x) = 1$, $e_1(x) = x$, and $e_2(x) = x^2$.*

Theorem II *Let (T_n) be a sequence of positive linear operators from $C_{2\pi}([0, 1])$ into $F([0, 1])$. Then $\lim_n \|T_n(f, x) - f(x)\|_\infty = 0$ for all $f \in C_{2\pi}([0, 1])$ if and only*

M. Mursaleen, S.A. Mohiuddine, *Convergence Methods for Double Sequences and Applications*, DOI 10.1007/978-81-322-1611-7_8, © Springer India 2014

if $\lim_n \|T_n(f_i, x) - f_i(x)\|_\infty = 0$ *for* $i = 0, 1, 2$, *where* $f_0(x) = 1$, $f_1(x) = \cos x$, *and* $f_2(x) = \sin x$.

Several mathematicians have worked on extending or generalizing the Korovkin theorems in many ways and to several settings, including function spaces, abstract Banach lattices, Banach algebras, Banach spaces, and so on. This theory is very useful in real analysis, functional analysis, harmonic analysis, measure theory, probability theory, summability theory and partial differential equations. But the foremost applications are concerned with constructive approximation theory, which uses it as a valuable tool. Even today, the development of Korovkin-type approximation theory is far from complete. Note that the first and second theorems of Korovkin are actually equivalent to the algebraic and trigonometric versions, respectively, of the classical Weierstrass approximation theorem [5]. For some recent work on this topic, we refer to [76].

8.2 Korovkin-Type Theorem via Statistical A-Summability

By $C(K)$ we denote the space of all continuous real-valued functions on any compact subset of the real two-dimensional space. Then $C(K)$ is a Banach space with the norm $\| \cdot \|_{C(K)}$ defined as

$$\|f\|_{C(K)} := \sup_{(x,y)\in K} |f(x, y)| \quad (f \in C(K)).$$

Before proceeding further, we recall the classical and statistical forms of Korovkin-type theorems studied in [37] and [124].

Theorem 8.1 [124] *Let* $\{L_{ij}\}$ *be a double sequence of positive linear operators acting from* $C(K)$ *into itself. Then, for all* $f \in C(K)$,

$$P\text{-}\lim_{m,n} \|L_{ij}(f) - f\|_{C(K)} = 0$$

if and only if

$$P\text{-}\lim_{m,n} \|L_{ij}(f_r) - f_r\|_{C(K)} = 0 \quad (r = 0, 1, 2, 3),$$

where $f_0(x, y) = 1$, $f_1(x, y) = x$, $f_2(x, y) = y$, $f_3(x, y) = x^2 + y^2$.

Theorem 8.2 [37] *Let* $A = (a_{ij}^{mn})$ *be a nonnegative RH-regular summability matrix. Let* $\{L_{ij}\}$ *be a double sequence of positive linear operators acting from* $C(K)$ *into itself. Then, for all* $f \in C(K)$,

$$\mathcal{S}_A\text{-}\lim_{m,n} \|L_{ij}(f) - f\|_{C(K)} = 0$$

if and only if

$$\mathcal{S}_A\text{-}\lim_{m,n}\left\|L_{ij}(f_r) - f_r\right\|_{C(K)} = 0 \quad (r = 0, 1, 2, 3),$$

where $f_0(x, y) = 1$, $f_1(x, y) = x$, $f_2(x, y) = y$, $f_3(x, y) = x^2 + y^2$.

By using the concept of statistical A-summability for single sequences, Korovkin-type theorems are proved in [35] and [36]. Now, we prove the following.

Theorem 8.3   *Let $A = (a_{ij}^{mn})$ be a nonnegative RH-regular summability matrix method. Let $\{L_{ij}\}$ be a double sequence of positive linear operators acting from $C(K)$ into itself. Then, for all $f \in C(K)$,*

$$\mathcal{S}\text{-}\lim_{m,n}\left\|\sum_{i,j=1,1}^{\infty,\infty} a_{ij}^{mn} L_{ij}(f) - f\right\|_{C(K)} = 0 \tag{8.1}$$

if and only if

$$\mathcal{S}\text{-}\lim_{m,n}\left\|\sum_{i,j=1,1}^{\infty,\infty} a_{ij}^{mn} L_{ij}(f_r) - f_r\right\|_{C(K)} = 0 \quad (r = 0, 1, 2, 3) \tag{8.2}$$

where $f_0(x, y) = 1$, $f_1(x, y) = x$, $f_2(x, y) = y$, $f_3(x, y) = x^2 + y^2$.

Proof Condition (8.2) follows immediately from condition (8.1) since each $f_r \in C(K)$ $(r = 0, 1, 2, 3)$. Let us prove the converse. By the continuity of f on the compact set K, we can write $|f(x, y)| \leq M$, where $M = \|f\|_{C(K)}$. Also, since $f \in C(K)$, for every $\epsilon > 0$, there is a number $\delta > 0$ such that $|f(u, v) - f(x, y)| < \epsilon$ for all $(u, v) \in K$ satisfying $|u - x| < \delta$ and $|v - y| < \delta$. Hence, we get

$$|f(u, v) - f(x, y)| < \epsilon + \frac{2M}{\delta^2}\{(u - x)^2 + (v - y)^2\}. \tag{8.3}$$

Since $L_{i,j}$ is linear and positive, from (8.3) we obtain that, for any $m, n \in \mathbb{N}$,

$$\left|\sum_{i,j=1,1}^{\infty,\infty} a_{ij}^{mn} L_{ij}(f; x, y) - f(x, y)\right|$$

$$\leq \sum_{i,j=1,1}^{\infty,\infty} a_{ij}^{mn} L_{ij}\big(|f(u, v) - f(x, y)|; x, y\big)$$

$$+ |f(x, y)|\left|\sum_{i,j=1,1}^{\infty,\infty} a_{ij}^{mn} L_{ij}(f_0; x, y) - f_0(x, y)\right|$$

$$\leq \sum_{i,j=1,1}^{\infty,\infty} a_{ij}^{mn} L_{ij}\left(\epsilon + \frac{2M}{\delta^2}[(u-x)^2 + (v-y)^2]; x, y\right)$$

$$+ |f(x,y)| \left| \sum_{i,j=1,1}^{\infty,\infty} a_{ij}^{mn} L_{ij}(f_0; x, y) - f_0(x,y) \right|$$

$$\leq \epsilon + (\epsilon + M) \left| \sum_{i,j=1,1}^{\infty,\infty} a_{ij}^{mn} L_{ij}(f_0; x, y) - f_0 \right|$$

$$+ \frac{2M}{\delta^2} \left\{ \left| \sum_{i,j=1,1}^{\infty,\infty} a_{ij}^{mn} L_{ij}(f_3; x, y) - f_3(x,y) \right| \right.$$

$$+ 2|x| \left| \sum_{i,j=1,1}^{\infty,\infty} a_{ij}^{mn} L_{ij}(f_1; x, y) - f_1(x,y) \right|$$

$$+ 2|y| \left| \sum_{i,j=1,1}^{\infty,\infty} a_{ij}^{mn} L_{ij}(f_2; x, y) - f_2(x,y) \right|$$

$$+ \left(x^2 + y^2\right) \left| \sum_{i,j=1,1}^{\infty,\infty} a_{ij}^{mn} L_{ij}(f_0; x, y) - f_0(x,y) \right|$$

$$\leq \epsilon + \left(\epsilon + M + \frac{2M}{\delta^2}\left(C^2 + D^2\right)\right) \left| \sum_{i,j=1,1}^{\infty,\infty} a_{ij}^{mn} L_{ij}(f_0; x, y) - f_0(x,y) \right|$$

$$+ \frac{2M}{\delta^2} \left| \sum_{i,j=1,1}^{\infty,\infty} a_{ij}^{mn} L_{ij}(f_3; x, y) - f_3(x,y) \right|$$

$$+ \frac{4MC}{\delta^2} \left| \sum_{i,j=1,1}^{\infty,\infty} a_{ij}^{mn} L_{ij}(f_1; x, y) - f_1(x,y) \right|$$

$$+ \frac{4MD}{\delta^2} \left| \sum_{i,j=1,1}^{\infty,\infty} a_{ij}^{mn} L_{ij}(f_2; x, y) - f_2(x,y) \right|,$$

where $C := \max |x|$ and $D := \max |y|$. Taking the supremum over $(x, y) \in K$, we get

$$\left\| \sum_{i,j=1,1}^{\infty,\infty} a_{ij}^{mn} L_{ij}(f) - f \right\| \leq \epsilon + B \sum_{r=0}^{3} \left\| \sum_{i,j=1,1}^{\infty,\infty} a_{ij}^{mn} L_{ij}(f_r; x, y) - f_r(x, y) \right\|,$$

where

$$B := \max\left\{\epsilon + M + \frac{2M}{\delta^2}(C^2 + D^2), \frac{2M}{\delta^2}, \frac{4MC}{\delta^2}, \frac{4MD}{\delta^2}\right\}.$$

Now for a given $\sigma > 0$, choose $\epsilon > 0$ such that $\epsilon < \sigma$ and define

$$E := \left\{(m, n) \in \mathbb{N}^2 : \left\|\sum_{i,j=1,1}^{\infty,\infty} a_{ij}^{mn} L_{ij}(f; x, y) - f(x, y)\right\| \geq \sigma\right\},$$

$$E_r := \left\{(m, n) \in \mathbb{N}^2 : \left\|\sum_{i,j=1,1}^{\infty,\infty} a_{ij}^{mn} L_{ij}(f_r; x, y) - f_r(x, y)\right\| \geq \frac{\sigma - \epsilon}{4B}\right\},$$

$$r = 0, 1, 2, 3.$$

Then $E \subset \bigcup_{r=0}^{3} E_r$, and so $\delta_2(E) \leq \sum_{r=0}^{3} \delta_2(E_r)$. By considering this inequality and using (8.2) we obtain (8.1), which completes the proof. $\square$

Example 8.4 Now we will show that Theorem 8.3 is stronger than its classical and statistical forms. Let $A = (a_{ij}^{mn})$ be a four-dimensional Cesàro matrix, i.e.,

$$a_{ij}^{mn} = \begin{cases} 1/mn & \text{if } i \leq m \text{ and } j \leq n, \\ 0 & \text{otherwise,} \end{cases}$$

and let $x = (x_{ij})$ be defined as

$$x_{ij} = (-1)^j \quad \text{for all } i.$$

Then this sequence is neither P-convergent nor A-statistically convergent, but S-$\lim Ax = 0$.

Now, consider the Bernstein operators (see [119]) defined for $f \in C(K)$ by

$$B_{ij}(f; x, y) = \sum_{k=0}^{i} \sum_{l=0}^{j} f\left(\frac{k}{i}, \frac{l}{j}\right) C(i, k) x^k (1 - x)^{i-k} C(j, l) y^j (1 - y)^{j-l} \quad (8.4)$$

for $(x, y) \in K = [0, 1] \times [0, 1]$. By using these operators, define the following positive linear operators on $C(K)$:

$$L_{ij}(f; x, y) = (1 + x_{ij}) B_{ij}(f; x, y), \quad (x, y) \in K, f \in C(K). \quad (8.5)$$

Then observe that

$$L_{ij}(f_0; x, y) = (1 + x_{ij}) f_0(x, y),$$

$$L_{ij}(f_1; x, y) = (1 + x_{ij}) f_1(x, y),$$

$$L_{ij}(f_2; x, y) = (1 + x_{ij}) f_2(x, y),$$

$$L_{ij}(f_3; x, y) = (1 + x_{ij})\left(f_3(x, y) + \frac{x - x^2}{i} + \frac{y - y^2}{j}\right),$$

where $f_0(x, y) = 1$, $f_1(x, y) = x$, $f_2(x, y) = y$, $f_3(x, y) = x^2 + y^2$. Since $S\text{-}\lim Ax = 0$, we obtain

$$S\text{-}\lim_{m,n}\left\|\sum_{i,j=1,1}^{\infty,\infty} a_{ij}^{mn} L_{ij}(f_r) - f_r\right\|_{C(K)} = S\text{-}\lim_{m,n}\frac{1}{mn}\left\|\sum_{i,j=1,1}^{\infty,\infty} L_{ij}(f_r) - f_r\right\|_{C(K)} = 0$$

for $r = 0, 1, 2, 3$. Hence, by Theorem 8.3 we conclude that

$$S\text{-}\lim_{m,n}\left\|\sum_{i,j=1,1}^{\infty,\infty} a_{ij}^{mn} L_{ij}(f) - f\right\|_{C(K)} = 0$$

for any $f \in C(K)$.

However, since the P-limit and the statistical limit of the double sequence (x_{ij}) are not zero, it follows that, for $r = 0, 1, 2, 3$, $\|L_{ij}(f_i) - f_i\|_{C(K)}$ is neither P-convergent nor statistically convergent to zero. So, Theorems 8.1 and 8.2 do not work for our operators defined by (8.4).

8.3 Korovkin-Type Theorem via A-Statistical Convergence

Boyanov and Veselinov [17] have proved the Korovkin theorem on $C[0, \infty)$ by using the test functions 1, e^{-x}, e^{-2x}. In this section, we first extend the result of Boyanov and Veselinov for functions of two variables by using the notion of P-convergence and further generalize for A-statistical convergence.

Theorem 8.5 *Let $(T_{j,k})$ be a double sequence of positive linear operators from $C(I^2)$ into $C(I^2)$. Then, for all $f \in C(I^2)$,*

$$P\text{-}\lim_{j,k\to\infty}\left\|T_{j,k}(f; x, y) - f(x, y)\right\|_{\infty} = 0 \tag{8.6}$$

if and only if

$$P\text{-}\lim_{j,k\to\infty}\left\|T_{j,k}(1; x, y) - 1\right\|_{\infty} = 0, \tag{8.7}$$

$$P\text{-}\lim_{j,k\to\infty}\left\|T_{j,k}(e^{-s}; x, y) - e^{-x}\right\|_{\infty} = 0, \tag{8.8}$$

$$P\text{-}\lim_{j,k\to\infty}\left\|T_{j,k}(e^{-t}; x, y) - e^{-y}\right\|_{\infty} = 0, \tag{8.9}$$

$$P\text{-}\lim_{j,k\to\infty}\left\|T_{j,k}(e^{-2s} + e^{-2t}; x, y) - (e^{-2x} + e^{-2y})\right\|_{\infty} = 0. \tag{8.10}$$

Proof Since each of the functions $1, e^{-x}, e^{-y}, e^{-2x} + e^{-2y}$ belongs to $C(I^2)$, conditions (8.7)–(8.10) follow immediately from (8.6). Let $f \in C(I^2)$. There exists a constant M such that $|f(x, y)| \leq M$ for all $(x, y) \in I^2$, where $M = \|f\|_\infty$. Therefore,

$$|f(s, t) - f(x, y)| \leq 2M, \quad 0 \leq s, t, x, y < \infty. \tag{8.11}$$

It is easy to prove that for given $\varepsilon > 0$, there is $\delta > 0$ such that

$$|f(s, t) - f(x, y)| < \varepsilon \tag{8.12}$$

whenever $|e^{-s} - e^{-x}| < \delta$ and $|e^{-t} - e^{-y}| < \delta$ for all $(x, y) \in I^2$.

Using (8.11), (8.12) and putting $\psi_1 = \psi_1(s, x) = (e^{-s} - e^{-x})^2$ and $\psi_2 = \psi_2(t, y) = (e^{-t} - e^{-y})^2$, we get

$$|f(s, t) - f(x, y)| < \varepsilon + \frac{2M}{\delta^2}(\psi_1 + \psi_2) \quad \forall |s - x| < \delta \text{ and } |t - y| < \delta,$$

that is,

$$-\varepsilon - \frac{2M}{\delta^2}(\psi_1 + \psi_2) < f(s, t) - f(x, y) < \varepsilon + \frac{2M}{\delta^2}(\psi_1 + \psi_2).$$

Now, operate $T_{j,k}(1; x, y)$ to this inequality. Since $T_{j,k}(f; x, y)$ is monotone and linear, we obtain

$$T_{j,k}(1; x, y)\left(-\varepsilon - \frac{2M}{\delta^2}(\psi_1 + \psi_2)\right) < T_{j,k}(1; x, y)\big(f(s, t) - f(x, y)\big)$$

$$< T_{j,k}(1; x, y)\left(\varepsilon + \frac{2M}{\delta^2}(\psi_1 + \psi_2)\right).$$

Note that x and y are fixed, and so $f(x, y)$ is a constant number. Therefore, by simple calculations we get

$$\big|T_{j,k}(f; x, y) - f(x, y)\big|$$

$$\leq \varepsilon + (\varepsilon + M)\big|T_{j,k}(1; x, y) - 1\big| + \frac{2M}{\delta^2}\big|e^{-2x} + e^{-2y}\big|\big|T_{j,k}(1; x, y) - 1\big|$$

$$+ \frac{2M}{\delta^2}\big|T_{j,k}(e^{-2s} + e^{-2t}; x, y) - (e^{-2x} + e^{-2y})\big|$$

$$+ \frac{4M}{\delta^2}\big|e^{-x}\big|\big|T_{j,k}(e^{-s}; x, y) - e^{-x}\big| + \frac{4M}{\delta^2}\big|e^{-y}\big|\big|T_{j,k}(e^{-t}; x, y) - e^{-y}\big|$$

$$\leq \varepsilon + \left(\varepsilon + M + \frac{4M}{\delta^2}\right)\big|T_{j,k}(1; x, y) - 1\big|$$

$$+ \frac{2M}{\delta^2}\big|T_{j,k}(e^{-2s} + e^{-2t}; x, y) - (e^{-2x} + e^{-2y})\big|$$

$$+ \frac{4M}{\delta^2} \left| T_{j,k}\left(e^{-s}; x, y\right) - e^{-x} \right| + \frac{4M}{\delta^2} \left| T_{j,k}\left(e^{-t}; x, y\right) - e^{-y} \right|. \tag{8.13}$$

Since $|e^{-x}|, |e^{-y}| \leq 1$ for all $x, y \in I$, taking $\sup_{(x,y) \in I^2}$, we get

$$\left\| T_{j,k}(f; x, y) - f(x, y) \right\|_\infty$$
$$\leq \varepsilon + K \left(\left\| T_{j,k}(1; x, t) - 1 \right\|_\infty + \left\| T_{j,k}\left(e^{-s}; x, y\right) - e^{-x} \right\|_\infty \right.$$
$$\left. + \left\| T_{j,k}\left(e^{-t}; x, y\right) - e^{-y} \right\|_\infty + \left\| T_{j,k}\left(e^{-2s} + e^{-2t}; x, y\right) - \left(e^{-2x} + e^{-2y}\right) \right\|_\infty \right),$$
$$\tag{8.14}$$

where $K = \max\{\varepsilon + M + \frac{4M}{\delta^2}, \frac{4M}{\delta^2}, \frac{2M}{\delta^2}\}$. Taking P-lim as $j, k \to \infty$ and using (8.7)–(8.10), we get

$$P\text{-} \lim_{p,q \to \infty} \left\| T_{j,k}(f; x, y) - f(x, y) \right\|_\infty = 0, \quad \text{uniformly in } m, n. \qquad \square$$

In the following theorem, we use the notion of almost convergence of double sequences to generalize the above theorem. We also give an example showing its importance.

Theorem 8.6 *Let $(T_{j,k})$ be a double sequence of positive linear operators from $C(I^2)$ into $C(I^2)$. Then, for all $f \in C(I^2)$,*

$$S_A\text{-} \lim_{j,k \to \infty} \left\| T_{j,k}(f; x, y) - f(x, y) \right\|_\infty = 0 \tag{8.15}$$

if and only if

$$S_A\text{-} \lim_{j,k \to \infty} \left\| T_{j,k}(1; x, y) - 1 \right\|_\infty = 0, \tag{8.16}$$

$$S_A\text{-} \lim_{j,k \to \infty} \left\| T_{j,k}\left(e^{-s}; x, y\right) - e^{-x} \right\|_\infty = 0, \tag{8.17}$$

$$S_A\text{-} \lim_{j,k \to \infty} \left\| T_{j,k}\left(e^{-t}; x, y\right) - e^{-y} \right\|_\infty = 0, \tag{8.18}$$

$$S_A\text{-} \lim_{j,k \to \infty} \left\| T_{j,k}\left(e^{-2s} + e^{-2t}; x, y\right) - \left(e^{-2x} + e^{-2y}\right) \right\|_\infty = 0. \tag{8.19}$$

Proof For a given $r > 0$, choose $\epsilon > 0$ such that $\epsilon < r$. Define the following sets:

$$D := \left\{ (j, k) \in \mathbb{N} \times \mathbb{N} : \left\| T_{j,k}(f; x, y) - f(x, y) \right\|_\infty \geq r \right\},$$

$$D_1 := \left\{ (j, k) \in \mathbb{N} \times \mathbb{N} : \left\| T_{j,k}(1; x, t) - 1 \right\|_\infty \geq \frac{r - \epsilon}{4K} \right\},$$

$$D_2 := \left\{ (j, k) \in \mathbb{N} \times \mathbb{N} : \left\| T_{j,k}\left(e^{-s}; x, y\right) - e^{-x} \right\|_\infty \geq \frac{r - \epsilon}{4K} \right\},$$

$$D_3 := \left\{ (j,k) \in \mathbb{N} \times \mathbb{N} : \left\| T_{j,k}\left(e^{-t}; x, y\right) - e^{-y} \right\|_\infty \geq \frac{r-\epsilon}{4K} \right\},$$

$$D_4 := \left\{ (j,k) \in \mathbb{N} \times \mathbb{N} : \left\| T_{j,k}\left(e^{-2s} + e^{-2t}; x, y\right) - \left(e^{-2x} + e^{-2y}\right) \right\|_\infty \geq \frac{r-\epsilon}{4K} \right\}.$$

Then by (8.14) it follows that $D \subset D_1 \cup D_2 \cup D_3 \cup D_4$. Hence, $\delta_A^{(2)}(D) \leq \delta_A^{(2)}(D_1) + \delta_A^{(2)}(D_2) + \delta_A^{(2)}(D_3) + \delta_A^{(2)}(D_4)$. Using (8.16)–(8.19), we get $\delta_A^{(2)}(D) = 0$, i.e.,

$$\mathcal{S}_A\text{-} \lim_{j,k \to \infty} \left\| T_{j,k}(f; x, y) - f(x, y) \right\|_\infty = 0.$$

$\square$

In the following example, we construct a double sequence of positive linear operators that satisfies the conditions of Theorem 8.6 but does not satisfy the conditions of Theorem 8.5, that is, Theorem 8.6 is stronger than Theorem 8.5.

Example 8.7 Consider the sequence of classical Baskakov operators of two variables [48]

$$B_{m,n}(f; x, y) := \sum_{j=0}^{\infty} \sum_{k=0}^{\infty} f\left(\frac{j}{m}, \frac{k}{n}\right) \binom{m-1+j}{j} \binom{n-1+k}{k}$$
$$\times x^j (1+x)^{-m-j} y^k (1+y)^{-n-k} \tag{8.20}$$

for $0 \leq x, y < \infty$.

Take A as in Example 8.4. Define a double sequence $z = (z_{mn})$ by

$$z_{mn} = \begin{cases} 1 & \text{if } m \text{ and } n \text{ are squares,} \\ 0 & \text{otherwise.} \end{cases}$$

Let $L_{m,n} : C(I^2) \to C(I^2)$ be defined by

$$L_{m,n}(f; x, y) = (1 + z_{mn}) B_{m,n}(f; x, y).$$

Since

$$B_{m,n}(1; x, y) = 1,$$
$$B_{m,n}\left(e^{-s}; x, y\right) = \left(1 + x - xe^{-\frac{1}{m}}\right)^{-m},$$
$$B_{m,n}\left(e^{-t}; x, y\right) = \left(1 + y - ye^{-\frac{1}{n}}\right)^{-n},$$
$$B_{m,n}\left(e^{-2s} + e^{-2t}; x, y\right) = \left(1 + x - xe^{-\frac{2}{m}}\right)^{-m} + \left(1 + y - ye^{-\frac{2}{n}}\right)^{-n},$$

we have that the sequence $(L_{m,n})$ satisfies conditions (8.16)–(8.19). Hence, by Theorem 8.6 we have

$$\mathcal{S}_A\text{-} \lim_{m,n \to \infty} \left\| L_{m,n}(f; x, y) - f(x, y) \right\|_\infty = 0.$$

On the other hand, we get $L_{m,n}(f; 0, 0) = (1 + z_{mn}) f(0, 0)$ since $B_{m,n}(f; 0, 0) = f(0, 0)$, and hence,

$$\left\| L_{m,n}(f; x, y) - f(x, y) \right\|_\infty \geq \left| L_{m,n}(f; 0, 0) - f(0, 0) \right| = z_{mn} \left| f(0, 0) \right|.$$

We see that $(L_{m,n})$ does not satisfy the conditions of Theorem 8.5 since P-$\lim_{m,n\to\infty} z_{mn}$ does not exist.

8.4 A-Statistical Approximation for Periodic Functions and Rate of A-Statistical Convergence

In this section, we present a Korovkin-type approximation theorem for periodic functions via A-statistical convergence and also study the rate of A-statistical convergence of a double sequence of positive linear operators defined from $C^*(\mathbb{R}^2)$ into $C^*(\mathbb{R}^2)$, where $C^*(\mathbb{R}^2)$ is the space of all 2π-periodic and real-valued continuous functions on $\mathbb{R}^2$ (see Demirci and Dirik [34] and Duman and Erkuş [38]).

Theorem 8.8 *Let $A = (a_{jkmn})$ be a nonnegative RH-regular summability matrix, and let (L_{mn}) be a double sequence of positive linear operators acting from $C^*(\mathbb{R}^2)$ into $C^*(\mathbb{R}^2)$. Then, for all $f \in C^*(\mathbb{R}^2)$,*

$$S_{A^-} \lim_{m,n\to\infty} \left\| L_{mn}(f) - f \right\|_{C^*(\mathbb{R}^2)} = 0 \tag{8.21}$$

if and only if

$$S_{A^-} \lim_{m,n\to\infty} \left\| L_{mn}(f_i) - f_i \right\|_{C^*(\mathbb{R}^2)} = 0, \quad i = 0, 1, 2, 3, 4, \tag{8.22}$$

where $f_0(x, y) = 1$, $f_1(x, y) = \sin x$, $f_2(x, y) = \sin y$, $f_3(x, y) = \cos x$, and $f_4(x, y) = \cos y$.

Proof Since each of the functions f_0, f_1, f_2, f_3, f_4 belongs to $C^*(\mathbb{R}^2)$, the necessity follows immediately from (8.21). Let conditions (8.22) hold, and let $f \in C^*(\mathbb{R}^2)$. Let I and J be closed intervals of length 2π. Fix $(x, y) \in I \times J$. By the continuity of f at (x, y) it follows that for given $\varepsilon > 0$, there is a number $\delta > 0$ such that, for all $(u, v) \in \mathbb{R}^2$,

$$\left| f(u, v) - f(x, y) \right| < \varepsilon \tag{8.23}$$

whenever $|u - x|, |v - y| < \delta$. Since f is bounded, it follows that

$$\left| f(u, v) - f(x, y) \right| \leq M_f = \| f \|_{C^*(\mathbb{R}^2)} \tag{8.24}$$

for all $(u, v) \in \mathbb{R}^2$.

For all $(u, v) \in (x - \delta, 2\pi + x - \delta] \times (y - \delta, 2\pi + y - \delta]$, it is well known that

$$|f(u, v) - f(x, y)| < \varepsilon + \frac{2M_f}{\sin^2 \frac{\delta}{2}} \psi(u, v), \qquad (8.25)$$

where $\psi(u, v) = \sin^2(\frac{u-x}{2}) + \sin^2(\frac{v-y}{2})$. Since the function $f \in C^*(\mathbb{R}^2)$ is 2π-periodic, inequality (8.25) holds for $(u, v) \in \mathbb{R}^2$. Then, we obtain

$$|L_{mn}(f; x, y) - f(x, y)|$$

$$\leq L_{mn}(|f(u, v) - f(x, y)|; x, y) + |f(x, y)||L_{mn}(f_0; x, y) - f_0(x, y)|$$

$$\leq \left| L_{mn}\left(\varepsilon + \frac{2M_f}{\sin^2 \frac{\delta}{2}} \psi(u, v); x, y\right) \right| + M_f |L_{mn}(f_0; x, y) - f_0(x, y)|$$

$$\leq \varepsilon + (\varepsilon + M_f)|L_{mn}(f_0; x, y) - f_0(x, y)| + \frac{M_f}{\sin^2 \frac{\delta}{2}} \{2|L_{mn}(f_0; x, y) - f_0(x, y)|$$

$$+ |\sin x||L_{mn}(f_1; x, y) - f_1(x, y)| + |\sin y||L_{mn}(f_2; x, y) - f_3(x, y)|$$

$$+ |\cos x||L_{mn}(f_3; x, y) - f_3(x, y)| + |\cos y||L_{mn}(f_4; x, y) - f_4(x, y)|\}$$

$$< \varepsilon + K \sum_{i=0}^{4} |L_{mn}(f_i; x, y) - f_i(x, y)|, \qquad (8.26)$$

where $K := \varepsilon + M_f + \frac{2M_f}{\sin^2 \frac{\delta}{2}}$. Now, taking $\sup_{(x,y) \in I \times J}$, we get

$$\|L_{mn}(f) - f\|_{C^*(\mathbb{R}^2)} < \varepsilon + K \sum_{i=0}^{4} \|L_{mn}(f_i) - f_i\|_{C^*(\mathbb{R}^2)}. \qquad (8.27)$$

Now for a given $r > 0$, choose $\varepsilon' > 0$ such that $\varepsilon' < r$. Define the following sets:

$$D = \left\{(m, n) : \|L_{mn}(f) - f\|_{C^*(\mathbb{R}^2)} \geq r\right\},$$

$$D_i = \left\{(m, n) : \|L_{mn}(f_i) - f_i\|_{C^*(\mathbb{R}^2)} \geq \frac{r - \varepsilon'}{5K}\right\},$$

where $i = 0, 1, 2, 3, 4$. Then, by (8.27),

$$D \subseteq \bigcup_{i=0}^{4} D_i,$$

and so

$$\sum_{(m,n) \in D} a_{jkmn} \leq \sum_{i=0}^{4} \sum_{(m,n) \in D_i} a_{jkmn},$$

i.e.,

$$\delta_A^{(2)}(D) \leq \sum_{i=0}^{4} \delta_A^{(2)}(D_i).$$

Now, using (8.22), we get

$$S_A\text{-}\lim_{m,n\to\infty} \left\| L_{mn}(f) - f \right\|_{C^*(\mathbb{R}^2)} = 0. \qquad\qquad \square$$

Remark 8.9 If we replace the matrix A by the identity matrix for four-dimensional matrices in Theorem 8.8, then we immediately get the following result in Pringsheim's sense.

Corollary 8.10 *Let $A = (a_{jkmn})$ be a nonnegative RH-regular summability matrix, and let (L_{mn}) be a double sequence of positive linear operators acting from $C^*(\mathbb{R}^2)$ into $C^*(\mathbb{R}^2)$. Then, for all $f \in C^*(\mathbb{R}^2)$,*

$$P\text{-}\lim_{m,n\to\infty} \left\| L_{mn}(f) - f \right\|_{C^*(\mathbb{R}^2)} = 0 \qquad\qquad (8.28)$$

if and only if

$$P\text{-}\lim_{m,n\to\infty} \left\| L_{mn}(f_i) - f_i \right\|_{C^*(\mathbb{R}^2)} = 0, \quad i = 0, 1, 2, 3, 4. \qquad (8.29)$$

Example 8.11 Now we present an example of double sequences of positive linear operators, showing that Corollary 8.10 does not work but our approximation theorem works. We consider the double sequence of Fejér operators on $C^*(\mathbb{R}^2)$

$$\sigma_{mn}(f; x, y) = \frac{1}{(n\pi)} \cdot \frac{1}{(n\pi)} \int_{-\pi}^{\pi} \int_{-\pi}^{\pi} f(u, v) F_m(u) F_n(v)\, du\, dv, \qquad (8.30)$$

where

$$F_m(u) = \frac{\sin^2(m(u - x)/2)}{\sin^2((u - x)/2)} \quad \text{and} \quad \frac{1}{\pi} \int_{-\pi}^{\pi} F_m(u)\, du = 1.$$

Observe that

$$\sigma_{mn}(f_0; x, y) = f_0(x, y), \qquad \sigma_{mn}(f_1; x, y) = \frac{m - 1}{m} f_1(x, y),$$

$$\sigma_{mn}(f_2; x, y) = \frac{n - 1}{n} f_2(x, y), \qquad \sigma_{mn}(f_3; x, y) = \frac{m - 1}{m} f_3(x, y), \qquad (8.31)$$

$$\text{and} \quad \sigma_{mn}(f_4; x, y) = \frac{n - 1}{n} f_4(x, y).$$

Now take $A = (C, 1, 1)$ and define the double sequence $\alpha = (\alpha_{mn})$ by

$$\alpha_{mn} = \begin{cases} 1 & \text{if } m \text{ and } n \text{ are squares,} \\ 0 & \text{otherwise.} \end{cases}$$

We observe that $\alpha = (\alpha_{mn})$ is not P-convergent but

$$\mathcal{S}_{(C,1,1)}\text{-}\lim \alpha = 0. \tag{8.32}$$

Let us define the operators $L_{mn} : C^*(\mathbb{R}^2) \rightarrow C^*(\mathbb{R}^2)$ by

$$L_{mn}(f; x, y) = (1 + \alpha_{mn})\sigma_{mn}(f; x, y). \tag{8.33}$$

Then, observe that the double sequence of positive linear operators (L_{mn}) defined by (8.33) satisfies all hypotheses of Theorem 8.8. Hence, by (8.31) we have that, for all $f \in C^*(\mathbb{R}^2)$,

$$\mathcal{S}_A\text{-}\lim_{m,n\to\infty} \|L_{mn}(f) - f\|_{C^*(\mathbb{R}^2)} = 0.$$

Since (α_{mn}) is not P-convergent, the sequence (L_{mn}) given by (8.33) does not converge uniformly to the function $f \in C^*(\mathbb{R}^2)$. So, we conclude that Corollary 8.10 does not work for the operators (L_{mn}) given by (8.33) while Theorem 8.8 still works. Hence, we conclude that the $\mathcal{S}_A$-version is stronger than the P-version.

Definition 8.12 Let $A = (a_{jkmn})$ be a nonnegative RH-regular summability matrix. Let (β_{mn}) be a positive nonincreasing double sequence. We say that a double sequence $x = (x_{mn})$ is A-statistically convergent to the number L with the rate $o(\beta_{mn})$ if for every $\varepsilon > 0$,

$$P\text{-}\lim_{j,k\to\infty} \frac{1}{\beta_{jk}} \sum_{(m,n)\in K(\epsilon)} a_{jkmn} = 0,$$

where $K(\epsilon) := \{(m, n) \in \mathbb{N} \times \mathbb{N} : |x_{mn} - \ell| \geq \epsilon\}$. In this case, we write $x_{mn} - L = st_A^{(2)} - o(\beta_{mn})$ as $m, n \to \infty$.

Now, we recall the notion of modulus of continuity. The modulus of continuity of $f \in C^*(\mathbb{R}^2)$, denoted by $\omega(f, \delta)$ for $\delta > 0$, is defined by

$$\omega(f, \delta) = \sup\{|f(u, v) - f(x, y)| : (u, v), (x, y) \in \mathbb{R}^2, \sqrt{(u-x)^2 + (v-y)^2} \leq \delta\}.$$

It is well known that

$$|f(u, v) - f(x, y)| \leq \omega\left(f, \sqrt{(u-x)^2 + (v-y)^2}\right)$$

$$\leq \omega(f, \delta)\left(\frac{\sqrt{(u-x)^2 + (v-y)^2}}{\delta} + 1\right). \tag{8.34}$$

Then we have the following result.

Theorem 8.13 *Let $A = (a_{jkmn})$ be a nonnegative RH-regular summability matrix, and let (L_{mn}) be a double sequence of positive linear operators acting from $C^*(\mathbb{R}^2)$ into $C^*(\mathbb{R}^2)$. Let (α_{mn}) and (β_{mn}) be two positive nonincreasing sequences. Suppose that*

(i) $\|L_{mn}(f_0) - f_0\|_{C^*(\mathbb{R}^2)} = \mathcal{S}_A\text{-}o(\alpha_{mn})$,

(ii) $\omega(f, \lambda_{mn}) = \mathcal{S}_A\text{-}o(\beta_{mn})$, where $\lambda_{mn} = \sqrt{\|L_{mn}(\varphi)\|_{C^*(\mathbb{R}^2)}}$ with

$$\varphi(u, v) = \sin^2\left(\frac{u - x}{2}\right) + \sin^2\left(\frac{v - y}{2}\right) \quad \text{for } (u, v), (x, y) \in \mathbb{R}^2.$$

Then, for all $f \in C^*(\mathbb{R}^2)$,

$$\|L_{mn}(f) - f\|_{C^*(\mathbb{R}^2)} = \mathcal{S}_A\text{-}o(\gamma_{mn}), \qquad (8.35)$$

where $\gamma_{mn} = \max\{\alpha_{mn}, \beta_{mn}\}$.

Proof Let $f \in C^*(\mathbb{R}^2)$ and $(x, y) \in [-\pi, \pi] \times [-\pi, \pi]$. Let $\delta > 0$. We have following cases.

Case I. If $\delta < |u - x| \leq \pi$, $\delta < |v - y| \leq \pi$, then $|u - x| \leq \pi |\sin \frac{u-x}{2}|$ and $|v - y| \leq \pi |\sin \frac{v-y}{2}|$. Therefore, by (8.34) we have

$$|f(u, v) - f(x, y)| \leq \omega(f, \delta)\left(\pi^2 \frac{\sin^2(\frac{u-x}{2}) + \sin^2(\frac{v-y}{2})}{\delta^2} + 1\right). \qquad (8.36)$$

Case II. $|u - x| > \pi$, $|v - y| \leq \pi$. Let k be an integer such that $|u + 2k\pi - x| \leq \pi$. Then

$$|f(u, v) - f(x, y)| = |f(u + 2k\pi, v) - f(x, y)|$$

$$\leq \omega(f, \delta)\left(\pi^2 \frac{\sin^2(\frac{u+2k\pi-x}{2}) + \sin^2(\frac{v-y}{2})}{\delta^2} + 1\right)$$

$$= \omega(f, \delta)\left(\pi^2 \frac{\sin^2(\frac{u-x}{2}) + \sin^2(\frac{v-y}{2})}{\delta^2} + 1\right).$$

Similarly, in other two cases where $|u - x| \leq \pi$, $|v - y| > \pi$ and $|u - x| > \pi$, $|v - y| > \pi$, we obtain (8.36).

Now, using the definition of modulus of continuity and the linearity and positivity of the operators L_{mn}, we get

$$|L_{mn}(f; x, y) - f(x, y)|$$

$$\leq L_{mn}\left(|f(u, v) - f(x, y)|; x, y\right) + |f(x, y)||L_{mn}(f_0; x, y) - f_0(x, y)|$$

$$\leq \omega(f, \delta)L_{mn}(f_0; x, y) + \pi^2 \frac{\omega(f, \delta)}{\delta^2} L_{mn}(\varphi; x, y)$$

$$+ |f(x, y)||L_{mn}(f_0; x, y) - f_0(x, y)|.$$

Taking the supremum over (x, y) on both sides of the above inequality and

$$\delta := \delta_{mn} = \sqrt{\|L_{mn}(\varphi)\|_{C^*(\mathbb{R}^2)}},$$

we obtain

$$\left\|L_{mn}(f)-f\right\|_{C^*(\mathbb{R}^2)} \le \omega(f,\delta_{mn})\left\|L_{mn}(f_0)-f_0\right\|_{C^*(\mathbb{R}^2)}$$
$$+ (1+\pi^2)\omega(f,\delta_{mn}) + M\left\|L_{mn}(f_0)-f_0\right\|_{C^*(\mathbb{R}^2)},$$
$$(8.37)$$

where $M := \|f\|_{C^*(\mathbb{R}^2)}$. Now, for a given $\varepsilon > 0$, define the following sets:

$$D = \left\{(m,n) : \left\|L_{mn}(f)-f\right\|_{C^*(\mathbb{R}^2)} \ge \varepsilon\right\},$$

$$D_1 = \left\{(m,n) : \left\|L_{mn}(f_0)-f_0\right\|_{C^*(\mathbb{R}^2)} \ge \frac{\varepsilon}{3}\right\},$$

$$D_2 = \left\{(m,n) : \omega(f,\delta_{mn}) \ge \frac{\varepsilon}{3(1+\pi^2)}\right\},$$

$$D_3 = \left\{(m,n) : \left\|L_{mn}(f)-f\right\|_{C^*(\mathbb{R}^2)} \ge \frac{\varepsilon}{3M}\right\}.$$

Then $D \subset D_1 \cup D_2 \cup D_3$. Further, defining

$$D_4 = \left\{(m,n) : \omega(f,\delta_{mn}) \ge \sqrt{\frac{\varepsilon}{3}}\right\},$$

$$D_5 = \left\{(m,n) : \left\|L_{mn}(f)-f\right\|_{C^*(\mathbb{R}^2)} \ge \sqrt{\frac{\varepsilon}{3}}\right\},$$

we see that $D_1 \subset D_4 \cup D_5$. Therefore, $D \subset \bigcup_{i=2}^5 D_i$. Therefore, since $\gamma_{mn} = \max\{\alpha_{mn}, \beta_{mn}\}$, we conclude that, for every $(j,k) \in \mathbb{N} \times \mathbb{N}$,

$$\frac{1}{\gamma_{jk}} \sum_{(m,n)\in D} a_{mnjk} \le \frac{1}{\alpha_{jk}} \sum_{(m,n)\in D_1} a_{mnjk} + \frac{1}{\beta_{jk}} \sum_{(m,n)\in D_2} a_{mnjk}$$

$$+ \frac{1}{\alpha_{jk}} \sum_{(m,n)\in D_3} a_{mnjk} + \frac{1}{\beta_{jk}} \sum_{(m,n)\in D_4} a_{mnjk}.$$

Letting $j,k \to \infty$ and using conditions (i) and (ii), we get

$$\left\|L_{mn}(f)-f\right\|_{C^*(\mathbb{R}^2)} = \mathcal{S}_A\text{-}o(\gamma_{mn}). \qquad \square$$

8.5 Exercises

1 Prove a Korovkin-type approximation theorem via A-statistical convergence of double sequences by using the test functions $1, x, y, x^2 + y^2$.

2 Prove Theorem 8.3 by using the test functions $1, \frac{x}{1-x}, \frac{y}{1-y}, (\frac{x}{1-x})^2 + (\frac{y}{1-y})^2$.

3 Prove Theorem 8.3 by using the test functions $1, \frac{x}{1+x}, \frac{y}{1+y}, (\frac{x}{1+x})^2 + (\frac{y}{1+y})^2$.

4 Prove Theorem 8.5 via statistical A-summability of double sequences.

5 Prove Theorem 8.6 via statistical A-summability of double sequences.

6 Prove Theorem 8.8 via statistical A-summability of double sequences.

Chapter 9
Double Series and Convergence Tests

Since Pringsheim introduced the notion of convergence of a numerical double series in terms of the convergence of the double sequence of its rectangular partial sums. However, an exhaustive treatment giving analogues of all well-known convergence aspects of single series seems to be unavailable. In this chapter, we give results to fill in some of the gaps in such a treatment and also to point out some errors in previous attempts to obtain results exactly analogous to those of a single series.

9.1 Introduction

In this chapter, we give some tests for absolute convergence of a double series including analogues of Cauchy's condensation test, Abel's kth term test, limit comparison test, ratio test, ratio comparison test, and Raabe's test. We give necessary and sufficient conditions on a double sequence $(a_{k\ell})$ in order that the Cauchy product double series $\sum_{k,\ell} a_{k\ell} b_{k\ell}$ would be convergent, boundedly convergent, or regularly convergent whenever a double series $\sum_{k,\ell} b_{k\ell}$ is convergent, boundedly convergent, or regularly convergent, respectively. We also show that if two double series are boundedly convergent, then the Cauchy product double series is Cesàro summable, and its Cesàro sum is equal to the product of the sums of the given double series. We compare these results with those obtained previously and give several examples to which these results apply. Although we shall consider, for simplicity, only double series whose terms are real numbers, the treatment carries over to multiple series whose terms may be complex numbers. Contents of this chapter are obtained from the paper of [62].

We shall use the partial order on $\mathbb{N}^2$ given by $(k_1, \ell_1) \leq (k_2, \ell_2)$ if and only if $k_1 \leq k_2$ and $\ell_1 \leq \ell_2$. The monotonicity of a double sequence is defined in terms of this partial order. We shall adopt Pringsheim's definition of convergence of a double series $\sum_{k,\ell} a_{k\ell}$ of real numbers: If $A_{m,n} := \sum_{k=1}^{m} \sum_{\ell=1}^{n} a_{k\ell}$ for $(m, n) \in \mathbb{N}^2$, then the series $\sum_{k,\ell} a_{k\ell}$ is said to be convergent if the double sequence $(A_{m,n})$ of its partial sums is convergent in the sense of Pringsheim, that is, there is $A \in \mathbb{R}$

M. Mursaleen, S.A. Mohiuddine, *Convergence Methods for Double Sequences and Applications*, DOI 10.1007/978-81-322-1611-7_9, © Springer India 2014

such that for every $\epsilon > 0$, there is $(m_0, n_0) \in \mathbb{N}^2$ satisfying $(m, n) \geq (m_0, n_0) \Rightarrow$ $|A_{m,n} - A| < \epsilon$. When every $a_{k\ell}$ is nonnegative, $\sum_{k,\ell} a_{k\ell}$ is convergent if and only if $(A_{m,n})$ is bounded above. For each fixed $k \in \mathbb{N}$, the series $\sum_\ell a_{k\ell}$ is called a row-series, and for each fixed $\ell \in \mathbb{N}$, the series $\sum_k a_{k\ell}$ is called a column-series corresponding to the double series $\sum_{k,\ell} a_{k\ell}$.

If a double series is absolutely convergent, then evidently the corresponding row-series and the column-series are all absolutely convergent. However, the converse is not true, as can be seen by considering $\sum_{k,\ell} a_{k\ell}$ where $a_{k,k} := 1$ and $a_{k,\ell} := 0$ for $k \neq \ell$. The following result gives necessary and sufficient conditions for the absolute convergence of a double series.

Lemma 9.1 *A double series $\sum_{k,\ell} a_{k\ell}$ is absolutely convergent if and only if the following conditions hold*:

(i) *There are $(k_0, \ell_0) \in \mathbb{N} \times \mathbb{N}$ and $\alpha_0 > 0$ such that*

$$\sum_{k=k_0}^{m} \sum_{\ell=\ell_0}^{n} |a_{k\ell}| \leq \alpha_0 \quad \text{for all } (m, n) \geq (k_0, \ell_0).$$

(ii) *Each row-series and each column-series are absolutely convergent.*

9.2   Convergence Tests

We shall provide a variety of conditions each of which imply condition (i) of the above lemma. These yield convergence tests for double series that are analogous to well-known convergence tests for single series.

9.2.1   Cauchy Condensation Test

The following test shows that we can study the convergence of certain double series by considering only some of its terms.

Theorem 9.2 (Cauchy condensation test) *Let $(a_{k,\ell})$ be a monotonically decreasing double sequence of nonnegative numbers. Then $\sum_{k,\ell=1}^{\infty} a_{k,\ell}$ converges if and only if $\sum_{k,\ell=1}^{\infty} 2^{k+\ell} a_{2^k, 2^\ell}$ converges.*

Proof Given $(m, n) \in \mathbb{N}^2$, let $i, j \in \mathbb{N}_0$ be such that $2^i \leq m < 2^{i+1}$ and $2^j \leq n < 2^{j+1}$. Since $a_{k,\ell} \geq 0$ for all $(k, \ell) \in \mathbb{N}^2$, we have

$$\sum_{k=0}^{i-1} \sum_{\ell=0}^{j-1} \left(\sum_{u=2^k}^{2^{k+1}-1} \sum_{v=2^\ell}^{2^{\ell+1}-1} a_{u,v} \right) \leq \sum_{k=1}^{m} \sum_{\ell=1}^{n} a_{k,\ell} \leq \sum_{k=0}^{i} \sum_{\ell=0}^{j} \left(\sum_{u=2^k}^{2^{k+1}-1} \sum_{v=2^\ell}^{2^{\ell+1}-1} a_{u,v} \right),$$

and since $(a_{k,\ell})$ is monotonically decreasing, we obtain

$$\frac{1}{4}\sum_{k=1}^{i}\sum_{\ell=1}^{j}2^{k+\ell}a_{2^k,2^\ell} = \sum_{k=0}^{i-1}\sum_{\ell=0}^{j-1}2^{k+\ell}a_{2^{k+1},2^{\ell+1}} \le \sum_{k=1}^{m}\sum_{\ell=1}^{n}a_{k,\ell} \le \sum_{k=0}^{i}\sum_{\ell=0}^{j}2^{k+\ell}a_{2^k,2^\ell}.$$

This shows that if the partial sums of $\sum_{k,\ell=0}^{\infty}2^{k+\ell}a_{2^k,2^\ell}$ are bounded, then so are the partial sums of $\sum_{k,\ell=1}^{\infty}a_{k,\ell}$, and if the partial sums of $\sum_{k,\ell=1}^{\infty}a_{k,\ell}$ are bounded, then so are the partial sums $\sum_{k,\ell=1}^{\infty}2^{k+\ell}a_{2^k,2^\ell}$. Further, if $\sum_{k,\ell=1}^{\infty}a_{k,\ell}$ is convergent, then the row-series $\sum_{\ell=1}^{\infty}a_{1,\ell}$ and the column-series $\sum_{k=1}^{\infty}a_{k,1}$ are convergent, where $(a_{k,1})$ and $(a_{1,\ell})$ are monotonically decreasing sequences of nonnegative numbers. In this case, the series $\sum_{k=0}^{\infty}2^k a_{2^k,1}$ and $\sum_{\ell=0}^{\infty}2^\ell a_{1,2^\ell}$ are convergent by Cauchy's condensation test for single series. Hence, the desired result follows. □

Example 9.3 Let $p \in \mathbb{R}$ and $a_{k,\ell} := 1/(k+\ell)^p$ for $(k,\ell) \in \mathbb{N}^2$. By Theorem 9.2, $\sum_{k,\ell=1}^{\infty}a_{k,\ell}$ converges if and only if the double series $\sum_{k,\ell=0}^{\infty}b_{k,\ell}$ converges, where $b_{k,\ell} := 2^{k+\ell}/(2^k+2^\ell)^p$. If $p \le 2$, then

$$b_{k,k} = \frac{2^{2k}}{(2 \cdot 2^k)^p} = \frac{1}{2^p}2^{k(2-p)} \ge \frac{1}{2^p} \quad \text{for } k \in \mathbb{N},$$

and so the double series $\sum_{k,\ell=0}^{\infty}b_{k,\ell}$ diverges. If $p > 2$, then

$$b_{k,\ell} = \frac{2^{k+\ell}}{(2^k+2^\ell)^p} \le \frac{2^{k+\ell}}{2^p(2^{k+\ell})^{p/2}} = \frac{1}{2^p}\left(2^{(2-p)/2}\right)^{k+\ell} \quad \text{for } (k,\ell) \in \mathbb{N}^2,$$

and so the double series $\sum_{k,\ell=0}^{\infty}b_{k,\ell}$ converges. It follows that the double series $\sum_{k,\ell=1}^{\infty}1/(k+\ell)^p$ converges if and only if $p > 2$.

9.2.2 Abel's (k,ℓ)th Term Test

If $\sum_{k,\ell}a_{k,\ell}$ is convergent, then $a_{k,\ell} \to 0$ as $k,\ell \to \infty$. This (k,ℓ)th term test is useful for establishing the divergence of a double series. The following variant of this test is analogous to Abel's kth term test for a single series.

Theorem 9.4 (Abel's (k,ℓ)th term test) *Suppose that $(a_{k,\ell})$ is a monotonically decreasing double sequence of nonnegative numbers. If the double series $\sum_{k,\ell}a_{k,\ell}$ is convergent, then $k\ell\, a_{k,\ell} \to 0$ as $k,\ell \to \infty$.*

Proof Given $(k,\ell) \in \mathbb{N}^2$, let $i_k, j_\ell \in \mathbb{N}_0$ be such that $2^{i_k} \le k < 2^{i_k+1}$ and $2^{j_\ell} \le \ell < 2^{j_\ell+1}$, and note that

$$0 \le k\ell\, a_{k,\ell} \le 2^{i_k+1}2^{j_\ell+1}a_{2^{i_k},2^{j_\ell}} = 4 \cdot 2^{i_k}2^{j_\ell}a_{2^{i_k},2^{j_\ell}}.$$

By Theorem 9.2, $\sum_{i,j=0}^{\infty} 2^{i+j} a_{2^i,2^j}$ is convergent, and so $2^{i+j} a_{2^i,2^j} \to 0$ as $i, j \to \infty$. Hence, $k\ell\, a_{k,\ell} \to 0$ as $k, \ell \to \infty$. $\qquad\qquad\square$

Example 9.5

(i) Let $p, q \in \mathbb{R}$ satisfy $p > 0$, $q > 0$ and $(1/p) + (1/q) \ge 1$, and define $a_{k,\ell} := 1/(k^p + \ell^q)$ for $(k, \ell) \in \mathbb{N}^2$. For $k \in \mathbb{N}$ and $\ell = [k^{p/q}]$, the integer part of $k^{p/q}$, we have

$$k\ell\, a_{k,\ell} = \frac{k\ell}{k^p + \ell^q} > \frac{k(k^{p/q} - 1)}{k^p + (k^{p/q})^q} = \frac{1}{2} k^{1-p+(p/q)} \left(1 - k^{-p/q}\right),$$

which does not tend to 0 as $k \to \infty$ since $1 - p + (p/q) \ge 0$ and $p/q > 0$. Hence, by Theorem 9.4 the double series $\sum_{k,\ell} 1/(k^p + \ell^q)$ diverges.

(ii) The converse of Theorem 9.4 does not hold. Define $a_{k,\ell} := 1/k\ell(\ln k)(\ln \ell)$ for $(k, \ell) \in \mathbb{N}^2$. Then $(a_{k,\ell})$ is a decreasing double sequence of nonnegative numbers, and $k\ell\, a_{k,\ell} \to 0$ as $k, \ell \to \infty$. However,

$$\sum_{k=1}^{m} \sum_{\ell=1}^{n} a_{k,\ell} = \left(\sum_{k=1}^{m} \frac{1}{k(\ln k)} \right) \left(\sum_{\ell=1}^{n} \frac{1}{\ell(\ln \ell)} \right) \to \infty \quad \text{as } m, n \to \infty.$$

9.2.3 Limit Comparison Test

Theorem 9.6 (Limit comparison test) *Let $(a_{k,\ell})$ and $(b_{k,\ell})$ be double sequences such that $a_{k,\ell} > 0$ and $b_{k,\ell} > 0$ for all $(k, \ell) \in \mathbb{N}^2$, each row-series and each column-series corresponding to both $\sum_{k,\ell} a_{k,\ell}$ and $\sum_{k,\ell} b_{k,\ell}$ are convergent, and $\lim_{k,\ell} a_{k,\ell}/b_{k,\ell} = r$, where $r \in \mathbb{R}$ and $r \ne 0$. Then $\sum_{k,\ell} a_{k,\ell}$ is convergent if and only if $\sum_{k,\ell} b_{k,\ell}$ is convergent.*

Proof Let the double series $\sum_{k,\ell} b_{k,\ell}$ be convergent. Then there is $\beta > 0$ such that $\sum_{k=1}^{m} \sum_{\ell=1}^{n} b_{k,\ell} \le \beta$ for all $(m, n) \in \mathbb{N}^2$. Since $a_{k,\ell}/b_{k,\ell} \to r$ as $k, \ell \to \infty$, there is $(k_0, \ell_0) \in \mathbb{N}^2$ such that $a_{k,\ell} \le (r + 1)b_{k,\ell}$ for all $(k, \ell) \ge (k_0, \ell_0)$. Hence, for all $(m, n) \ge (k_0, \ell_0)$, we have

$$\sum_{k=k_0}^{m} \sum_{\ell=\ell_0}^{n} a_{k,\ell} \le (r + 1) \sum_{k=k_0}^{m} \sum_{\ell=\ell_0}^{n} b_{k,\ell} \le (r + 1)\beta.$$

By Lemma 9.1 it follows that the double series $\sum_{k,\ell} a_{k,\ell}$ is convergent.

Conversely, let the double series $\sum_{k,\ell} a_{k,\ell}$ be convergent. Since $\lim_{k,\ell} b_{k,\ell}/a_{k,\ell} = 1/r$, the convergence of the double series $\sum_{k,\ell} b_{k,\ell}$ follows from the first part of the proof by interchanging $a_{k,\ell}$ and $b_{k,\ell}$. $\qquad\square$

9.2.4 Ratio Test

We shall now develop several convergence tests involving ratios of 'consecutive' terms of a double series.

Theorem 9.7 (Ratio test) *Let* $(a_{k,\ell})$ *be a double sequence of nonzero numbers such that either* $|a_{k,\ell+1}|/|a_{k,\ell}| \to a$ *or* $|a_{k+1,\ell}|/|a_{k,\ell}| \to \tilde{a}$ *as* $k, \ell \to \infty$*, where* $a, \tilde{a}, \in \mathbb{R} \cup \{\infty\}$*.*

(i) *Suppose that each row-series and each column-series corresponding to* $\sum_{k,\ell} a_{k,\ell}$ *are absolutely convergent. If* $a < 1$ *or* $\tilde{a} < 1$*, then* $\sum_{k,\ell} a_{k,\ell}$ *is absolutely convergent.*

(ii) *If* $a > 1$ *or* $\tilde{a} > 1$*, then* $\sum_{k,\ell} a_{k,\ell}$ *is divergent.*

Proof (i) Assume that $a < 1$. Then there are $\alpha \in (0, 1)$ and $(k_0, \ell_0) \in \mathbb{N}^2$ such that $|a_{k,\ell+1}| \leq \alpha |a_{k,\ell}|$ for all $(k, \ell) \geq (k_0, \ell_0)$. Hence,

$$|a_{k,\ell}| \leq \alpha |a_{k,\ell-1}| \leq \cdots \leq \alpha^{\ell-\ell_0} |a_{k,\ell_0}| \quad \text{for all } (k, \ell) \geq (k_0, \ell_0 + 1).$$

Since $\alpha < 1$, we have $\sum_{\ell=1}^{n} \alpha^\ell \leq 1/(1 - \alpha)$ for all $n \in \mathbb{N}$. Also, since the series $\sum_k a_{k,\ell_0}$ is assumed to be absolutely convergent, there is $\beta > 0$ such that $\sum_{k=1}^{m} |a_{k,\ell_0}| \leq \beta$ for all $m \in \mathbb{N}$. Hence, we obtain

$$\sum_{k=k_0}^{m} \sum_{\ell=\ell_0+1}^{n} |a_{k,\ell}| \leq \frac{\alpha^{-\ell_0} \beta}{1 - \alpha} \quad \text{for all } (m, n) \geq (k_0, \ell_0 + 1).$$

By Lemma 9.1 it follows that $\sum_{k,\ell} a_{k,\ell}$ is absolutely convergent. A similar argument holds if $\tilde{a} < 1$ instead of $a < 1$.

(ii) Assume that $a \in \mathbb{R}$ with $a > 1$ or $a = \infty$. Then there are $\alpha \in (1, \infty)$ and $(k_0, \ell_0) \in \mathbb{N}^2$ such that $|a_{k,\ell+1}|/|a_{k,\ell}| \geq \alpha$ for all $(k, \ell) \geq (k_0, \ell_0)$. Hence,

$$|a_{k,\ell}| \geq \alpha |a_{k,\ell-1}| \geq \cdots \geq \alpha^{\ell-\ell_0} |a_{k,\ell_0}| > 0 \quad \text{for all } (k, \ell) \geq (k_0, \ell_0 + 1).$$

For a fixed $k \in \mathbb{N}$ with $k \geq k_0$, there is $\ell_1 \in \mathbb{N}$ such that $|a_{k,\ell}| \geq \alpha^{\ell-\ell_0} |a_{k,\ell_0}| \geq 1$ for all $\ell \geq \ell_1$. Hence, $a_{k,\ell} \nrightarrow 0$ as $k, \ell \to \infty$, so that $\sum_{k,\ell} a_{k,\ell}$ is divergent. The same conclusion holds if $\tilde{a} \in \mathbb{R}$ with $\tilde{a} > 1$ or $\tilde{a} = \infty$. $\qquad \square$

Remarks 9.8

(i) The proof of part (ii) of Theorem 9.7 shows that if $a > 1$, then the row-series $\sum_\ell a_{k,\ell}$ diverges for each (fixed) large k. Hence, if each row-series converges and the limit a exists, then $a \leq 1$. Similarly, if $\tilde{a} > 1$, then the column-series $\sum_k a_{k,\ell}$ diverges for each (fixed) large ℓ. Hence, if each column-series converges and the limit $\tilde{a}$ exists, then $\tilde{a} \leq 1$.

(ii) Suppose that each row-series and each column-series corresponding to $\sum_{k,\ell} a_{k,\ell}$ are absolutely convergent. If $a = \tilde{a} = 1$, then the double series may converge or may diverge, as Example 9.3 shows.

(iii) For a double sequence $(a_{k,\ell})$, consider the limits $b_k := \lim_\ell a_{k,\ell+1}/a_{k,\ell}$ for a fixed $k \in \mathbb{N}$ and $c_\ell := \lim_k a_{k+1,\ell}/a_{k,\ell}$ for a fixed $\ell \in \mathbb{N}$, whenever they exist. Biermann (page 123 of [14]) and Vorob'ev (Sect. 4 of Chap. 13 in [125]) claimed that if b_k exists and is less than 1 for each $k \in \mathbb{N}$ and if c_ℓ exists and is less than 1 for each $\ell \in \mathbb{N}$, then the double series $\sum_{k,\ell} a_{k,\ell}$ is absolutely convergent. Although the ratio test for single series shows that each row-series and each column-series corresponding to $\sum_{k,\ell} a_{k,\ell}$ are absolutely convergent, the double series $\sum_{k,\ell} a_{k,\ell}$ may not be absolutely convergent. For example, define $a_{k,\ell} := 1/(2^k 2^{\ell/2^k})$ for (k, ℓ) in $\mathbb{N}^2$. Then $b_k < 1$ for each $k \in \mathbb{N}$, and $c_\ell < 1$ for each $\ell \in \mathbb{N}$. However, since for each fixed $k \in \mathbb{N}$,

$$\sum_{\ell=1}^\infty a_{k,\ell} = \frac{1}{2^k} \sum_{\ell=1}^\infty \left(\frac{1}{2^{1/2^k}}\right)^\ell = \frac{1}{2^k(2^{1/2^k} - 1)},$$

and since $1/2^k(2^{1/2^k} - 1) \to 1/\ln 2$ as $k \to \infty$, we see that the iterated series $\sum_k \sum_\ell a_{k,\ell}$ diverges, and so the double series $\sum_{k,\ell} a_{k,\ell}$ also diverges. Thus, the claims of Biermann and Vorob'ev are incorrect.

(iv) Suppose that $a_{k,\ell} > 0$ for all $(k, \ell) \in \mathbb{N}^2$ and that each row-series and each column-series corresponding to $\sum_{k,\ell} a_{k,\ell}$ are convergent. A rather involved version of ratio test is given by Baron in Sect. 2 of [9] as follows. If $\lim_{k,\ell} a_{k,\ell}$ exists and if the limit

$$d := \lim_{k,\ell} \frac{a_{k+1,\ell} + a_{k,\ell+1} - a_{k+1,\ell+1}}{a_{k,\ell}}$$

exists with $d < 1$, then the double series $\sum_{k,\ell} a_{k,\ell}$ is convergent. Let us compare Baron's version of ratio test with our Theorem 9.7. Suppose that both the limits a and $\tilde{a}$ stated in Theorem 9.7 exist and are in $\mathbb{R}$. Then the limit d exists and

$$d = \lim_{k,\ell} \left(\frac{a_{k+1,\ell}}{a_{k,\ell}} + \frac{a_{k,\ell+1}}{a_{k,\ell}} - \frac{a_{k+1,\ell+1}}{a_{k+1,\ell}} \frac{a_{k+1,\ell}}{a_{k,\ell}}\right) = \tilde{a} + a - a\tilde{a}.$$

Since, in view of (i) above, we have $a \le 1$ and $\tilde{a} \le 1$, and since $1 - d = (1 - a)(1 - \tilde{a})$, we see that $d < 1$ if and only if $a < 1$ and $\tilde{a} < 1$. Thus, if one of a and $\tilde{a}$ is equal to 1 and the other is not, then $d = 1$, and hence Theorem 9.7 is applicable, but Baron's version of ratio test is not. For example, if $a_{k,\ell} := 1/k^2 2^\ell$ for $(k, \ell) \in \mathbb{N}^2$, then $a = 1/2$, $\tilde{a} = 1$, and $d = 1$.

Now suppose that the limit a exists and it is a real number other than 1. If the limit d exists, then it can be seen that the limit $\tilde{a}$ exists and is equal to $(d - a)/(1 - a)$. Thus, if $a < 1$ and $\tilde{a}$ does not exist, then the limit d cannot exist, and hence Theorem 9.7 is applicable, but Baron's version of ratio test is not. For example, if $a_{k,\ell} := 1/2^{(k^2+k\ell+\ell)/k}$, then $a = 1/2$, while the limits $\tilde{a}$ and d do not exist.

9.2.5 Ratio Comparison Test

Now we consider an analogue of the ratio comparison test for single series. (See, for example, Theorem 6 in Chap. 5 of [18].)

Theorem 9.9 (Ratio comparison test) *Let $(a_{k,\ell})$ and $(b_{k,\ell})$ be double sequences with $b_{k,\ell} > 0$ for all $(k, \ell) \in \mathbb{N}^2$.*

(i) *Suppose that each row-series and each column-series corresponding to $\sum_{k,\ell} |a_{k,\ell}|$ are convergent. If $|a_{k,\ell+1}| b_{k,\ell} \le |a_{k,\ell}| b_{k,\ell+1}$ and $|a_{k+1,\ell}| b_{k,\ell} \le |a_{k,\ell}| b_{k+1,\ell}$ whenever k and ℓ are large, and if $\sum_{k,\ell} b_{k,\ell}$ is convergent, then so is $\sum_{k,\ell} |a_{k,\ell}|$.*

(ii) *If $|a_{k,\ell+1}| b_{k,\ell} \ge |a_{k,\ell}| b_{k,\ell+1} > 0$ whenever ℓ is large and $k \in \mathbb{N}$, $|a_{k+1,\ell}| b_{k,\ell} \ge |a_{k,\ell}| b_{k+1,\ell} > 0$ whenever k is large and $\ell \in \mathbb{N}$, and if $\sum_{k,\ell} b_{k,\ell}$ is divergent, then so is $\sum_{k,\ell} |a_{k,\ell}|$.*

Proof (i) Let $k_0, \ell_0 \in \mathbb{N}$ be such that $|a_{k,\ell+1}| b_{k,\ell} \le |a_{k,\ell}| b_{k,\ell+1}$ and $|a_{k+1,\ell}| b_{k,\ell} \le |a_{k,\ell}| b_{k+1,\ell}$ for all $(k, \ell) \ge (k_0, \ell_0)$. Then

$$\frac{|a_{k,\ell}|}{b_{k,\ell}} \le \frac{|a_{k,\ell-1}|}{b_{k,\ell-1}} \le \cdots \le \frac{|a_{k,\ell_0}|}{b_{k,\ell_0}} \le \frac{|a_{k-1,\ell_0}|}{b_{k-1,\ell_0}} \le \cdots \le \frac{|a_{k_0,\ell_0}|}{b_{k_0,\ell_0}} \quad \text{for } (k, \ell) \ge (k_0, \ell_0).$$

Since $\beta := \sup\{\sum_{k=1}^m \sum_{\ell=1}^n b_{k,\ell} : (m, n) \in \mathbb{N}^2\} < \infty$, we obtain

$$\sum_{k=k_0}^m \sum_{\ell=\ell_0}^n |a_{k,\ell}| \le \frac{|a_{k_0,\ell_0}|}{b_{k_0,\ell_0}} \sum_{k=k_0}^m \sum_{\ell=\ell_0}^n b_{k,\ell} \le \beta \frac{|a_{k_0,\ell_0}|}{b_{k_0,\ell_0}} \quad \text{for all } m \ge k_0 \text{ and } n \ge \ell_0.$$

Hence, by Lemma 9.1 the double series $\sum_{k,\ell} |a_{k,\ell}|$ converges.

(ii) Suppose that $k_0 \in \mathbb{N}$ is such that $|a_{k+1,\ell}| b_{k,\ell} \ge |a_{k,\ell}| b_{k+1,\ell} > 0$ for $k \ge k_0$ and $\ell \in \mathbb{N}$ and that $\ell_0 \in \mathbb{N}$ is such that $|a_{k,\ell+1}| b_{k,\ell} \ge |a_{k,\ell}| b_{k,\ell+1} > 0$ for $\ell \ge \ell_0$ and $k \in \mathbb{N}$. If $\sum_k b_{k,\ell}$ diverges for some $\ell \in \mathbb{N}$, then by the ratio comparison test for single series, $\sum_k |a_{k,\ell}|$ also diverges for that ℓ. Similarly, if $\sum_\ell b_{k,\ell}$ diverges for some $k \in \mathbb{N}$, then $\sum_\ell |a_{k,\ell}|$ also diverges for that k. In these cases, condition (ii) of Lemma 9.1 is not satisfied, and so the double series $\sum_{k,\ell} |a_{k,\ell}|$ diverges. In the remaining case, the set $\{\sum_{k=k_0}^m \sum_{\ell=\ell_0}^n b_{k,\ell} : (m, n) \in \mathbb{N}^2\}$ is unbounded. Reversing the inequality signs in (i) above, we obtain

$$\sum_{k=k_0}^m \sum_{\ell=\ell_0}^n |a_{k,\ell}| \ge \frac{|a_{k_0,\ell_0}|}{b_{k_0,\ell_0}} \sum_{k=k_0}^m \sum_{\ell=\ell_0}^n b_{k,\ell},$$

which tends to ∞ as $m, n \to \infty$. Hence, by Lemma 9.1 $\sum_{k,\ell} |a_{k,\ell}|$ diverges. $\square$

Remarks 9.10

(i) The following example shows that both the inequalities $|a_{k,\ell+1}|b_{k,\ell} \leq |a_{k,\ell}|b_{k,\ell+1}$ and $|a_{k+1,\ell}|b_{k,\ell} \leq |a_{k,\ell}|b_{k+1,\ell}$ are needed in part (i) of Theorem 9.9. Define

$$a_{k,\ell} := \frac{1}{(k+\ell)^2} \quad \text{and} \quad b_{k,\ell} = \frac{1}{2^k(k+\ell)^2} \quad \text{for } (k,\ell) \in \mathbb{N}^2.$$

Although each row-series and each column-series converge, the double series $\sum_{k,\ell} a_{k,\ell}$ diverges, as we have seen in Example 9.3. However, the double series $\sum_{k,\ell} b_{k,\ell}$ converges since $1/(2^k(k+\ell)^2) \leq 1/(2^k \ell^2)$ for all $(k,\ell) \in \mathbb{N}^2$. Here the first inequality mentioned above holds, but the second does not. To see that both the inequalities $|a_{k,\ell+1}|b_{k,\ell} \geq |a_{k,\ell}|b_{k,\ell+1}$ and $|a_{k+1,\ell}|b_{k,\ell} \geq |a_{k,\ell}|b_{k+1,\ell}$ are needed in part (ii) of Theorem 9.9, we just interchange the roles of $a_{k,\ell}$ and $b_{k,\ell}$ in (i) above.

(ii) The requirement "$|a_{k,\ell+1}|b_{k,\ell} \leq |a_{k,\ell}|b_{k,\ell+1}$ and $|a_{k+1,\ell}|b_{k,\ell} \leq |a_{k,\ell}|b_{k+1,\ell}$ whenever k and ℓ are large" in part (i) of Theorem 9.9 is less stringent than the requirement "$|a_{k+p,\ell+1+q}|b_{k+r,\ell+s} \leq |a_{k+p,\ell+q}|b_{k+r,\ell+1+s}$ and $|a_{k+1+p,\ell+q}|b_{k+r,\ell+s} \leq |a_{k+1+p,\ell+q}|b_{k+1+r,\ell+s}$ for all $k,\ell,p,q,r,s \in \mathbb{N}$" imposed by Biermann for a similar result in [14, page 124].

9.2.6 Raabe's Test

As a consequence of the ratio comparison test, we obtain an analogue of Raabe's test for single series. It is useful in some cases when $a = \tilde{a} = 1$ in Theorem 9.7.

Theorem 9.11 (Raabe's test) *Let $(a_{k,\ell})$ be a double sequence.*

(i) *Suppose that each row-series and each column-series corresponding to $\sum_{k,\ell} |a_{k,\ell}|$ are convergent. If there is $p > 1$ such that*

$$|a_{k,\ell+1}| \leq \left(1 - \frac{p}{\ell}\right)|a_{k,\ell}| \quad \text{and} \quad |a_{k+1,\ell}| \leq \left(1 - \frac{p}{k}\right)|a_{k,\ell}|$$

whenever k and ℓ are large, then $\sum_{k,\ell} |a_{k,\ell}|$ is convergent.

(ii) *If $|a_{k,\ell+1}| \geq (1 - \frac{1}{\ell})|a_{k,\ell}| > 0$ for some $k \in \mathbb{N}$ and all large $\ell \in \mathbb{N}$, or if $|a_{k+1,\ell}| \geq (1 - \frac{1}{k})|a_{k,\ell}| > 0$ for some $\ell \in \mathbb{N}$ and all large $k \in \mathbb{N}$, then $\sum_{k,\ell} |a_{k,\ell}|$ is divergent.*

Proof (i) Suppose that there is $p > 1$ with the stated properties. Using the inequality $1 - px \leq (1-x)^p$ for $x \in [0,1]$, we obtain

$$|a_{k,\ell+1}| \leq \left(1 - \frac{p}{\ell}\right)|a_{k,\ell}| \leq \left(1 - \frac{1}{\ell}\right)^p |a_{k,\ell}| \leq \left(\frac{\ell}{\ell+1}\right)^p |a_{k,\ell}|$$

whenever k and ℓ are large, and hence

$$|a_{k,\ell+1}|\frac{1}{k^p\ell^p} \le |a_{k,\ell}|\frac{1}{k^p(\ell+1)^p}.$$

Similarly, we obtain

$$|a_{k+1,\ell}|\frac{1}{k^p\ell^p} \le |a_{k,\ell}|\frac{1}{(k+1)^p\ell^p}$$

whenever k and ℓ are large. By part (i) of Theorem 9.9 with $b_{k,\ell} := 1/(k^p\ell^p)$ for $(k,\ell) \in \mathbb{N}^2$, we obtain the desired result.

(ii) Suppose that the assumption in (ii) holds. Then by Raabe's test for single series, $\sum_k |a_{k,\ell}|$ diverges for some $\ell \in \mathbb{N}$ or $\sum_\ell |a_{k,\ell}|$ diverges for some $k \in \mathbb{N}$. In any case, condition (ii) of Lemma 9.1 is not satisfied, and so the double series $\sum_{k,\ell} |a_{k,\ell}|$ is divergent. □

Examples 9.12

(i) Let $a_{1,1} := 1$, $a_{k+1,\ell} := (k-1)a_{k,\ell}/(k+1)$ and $a_{k,\ell+1} := (\ell-1)a_{k,\ell}/(\ell+1)$ for $k, \ell \in \mathbb{N}$. By part (i) of Theorem 9.11 with $p = 3/2$ we see that the double series $\sum_{k,\ell} a_{k,\ell}$ is convergent.
(ii) Let $a_{1,1} := 1$, $a_{k+1,\ell} := ka_{k,\ell}/(k+1)$, and $a_{k,\ell+1} := \ell\, a_{k,\ell}/(\ell+1)$. By part (ii) of Theorem 9.11 we see that the double series $\sum_{k,\ell} a_{k,\ell}$ is divergent.

In both examples, Theorem 9.7 is not applicable since $a = \tilde{a} = 1$.

We deduce the following "limit" version of Raabe's test from Theorem 9.11.

Theorem 9.13 *Let $(a_{k,\ell})$ be a double sequence of nonzero numbers.*

(i) *Suppose that $\ell\,(1 - |a_{k,\ell+1}/a_{k,\ell}|) \to \alpha$ and $k\,(1 - |a_{k+1,\ell}/a_{k,\ell}|) \to \tilde{\alpha}$ as $k, \ell \to \infty$, where $\alpha, \tilde{\alpha}, \in \mathbb{R} \cup \{\infty\}$. Suppose that each row-series and each column-series corresponding to $\sum_{k,\ell} |a_{k,\ell}|$ are convergent. If $\alpha > 1$ and $\tilde{\alpha} > 1$, then $\sum_{k,\ell} |a_{k,\ell}|$ is convergent.*
(ii) *If for some $k \in \mathbb{N}$, the limit $\lim_{\ell\to\infty} \ell(1 - |\frac{a_{k,\ell+1}}{a_{k,\ell}}|)$ exists and is less than 1, or if for some $\ell \in \mathbb{N}$, the limit $\lim_{k\to\infty} k(1 - |\frac{a_{k+1,\ell}}{a_{k,\ell}}|)$ exists and is less than 1, then $\sum_{k,\ell} |a_{k,\ell}|$ is divergent.*

Remark 9.14 Suppose that $a_{k,\ell} > 0$ for all $(k, \ell) \in \mathbb{N}^2$ and that each row-series and each column-series corresponding to $\sum_{k,\ell} a_{k,\ell}$ are convergent. A rather involved version of Raabe's test for double series is given by Baron in Sect. 3 of [9] as follows. If $\lim_{k,\ell} a_{k,\ell}$ exists and if the limit

$$r := \lim_{k,\ell}\left[(k+\ell)\left(1 - \frac{a_{k+1,\ell} + a_{k,\ell+1} - a_{k+1,\ell+1}}{a_{k,\ell}}\right) + \frac{a_{k+1,\ell+1}}{a_{k,\ell}}\right]$$

exists with $r > 1$, then the double series $\sum_{k,\ell} a_{k,\ell}$ is convergent. Let us compare Baron's version of Raabe's with our Theorem 9.13. Suppose that both the limits α and $\tilde{\alpha}$ stated in Theorem 9.13 exist and are in $\mathbb{R}$. Then since $a_{k,\ell+1}/a_{k,\ell} \to 1$ and $a_{k+1,\ell}/a_{k,\ell} \to 1$ as $k, \ell \to \infty$, we see that the limit r exists and

$$
\begin{aligned}
r = \lim_{k,\ell} \Bigg[& k\left(1 - \frac{a_{k+1,\ell}}{a_{k,\ell}}\right) + \ell\left(1 - \frac{a_{k,\ell+1}}{a_{k,\ell}}\right) - \frac{a_{k+1,\ell}}{a_{k,\ell}}\ell\left(1 - \frac{a_{k+1,\ell+1}}{a_{k+1,\ell}}\right) \\
& - \frac{a_{k,\ell+1}}{a_{k,\ell}}k\left(1 - \frac{a_{k+1,\ell+1}}{a_{k,\ell+1}}\right) + \frac{a_{k+1,\ell+1}}{a_{k,\ell+1}}\frac{a_{k,\ell+1}}{a_{k,\ell}} \Bigg] \\
= & \, \tilde{\alpha} + \alpha - \alpha - \tilde{\alpha} + 1 = 1.
\end{aligned}
$$

In all such cases, Theorem 9.13 is applicable, but Baron's version of Raabe's test is not. For instance, in Example 9.12(i), we have $\alpha = 2 = \tilde{\alpha}$, but $r = 1$.

9.3 Cauchy Product

The Cauchy product of sequences (a_k) and (b_k) with $k \in \mathbb{N}_0$, is defined to be the sequence $(a_k * b_k)$, where $a_k * b_k := \sum_{i=0}^{k} a_i b_{k-i}$ for $k \in \mathbb{N}_0$, and the Cauchy product of single series $\sum_{k=0}^{\infty} a_k$ and $\sum_{k=0}^{\infty} b_k$ is defined to be the double series $\sum_{k=0}^{\infty} a_k * b_k$. Analogously, the *Cauchy product* of double sequences $(a_{k,\ell})$ and $(b_{k,\ell})$ with $(k, \ell) \in \mathbb{N}_0^2$, is defined to be the sequence $(a_{k,\ell} * b_{k,\ell})$ defined as

$$
a_{k,\ell} * b_{k,\ell} := \sum_{i=0}^{k} \sum_{j=0}^{\ell} a_{i,j} b_{k-i,\ell-j} \quad \text{for } (k, \ell) \in \mathbb{N}_0^2,
$$

and the Cauchy product of double series $\sum_{k,\ell=0}^{\infty} a_{k,\ell}$ and $\sum_{k,\ell=0}^{\infty} b_{k,\ell}$ is defined to be the double series $\sum_{k,\ell=0}^{\infty} a_{k,\ell} * b_{k,\ell}$. A classical result of Mertens states that if one of the given single series is absolutely convergent and the other is convergent, then their Cauchy product series is convergent. Another result due to Abel states that if both the given single series and their Cauchy product series are convergent, then the sum of the Cauchy product series is equal to the product of the sums of the given series. It has been known for long that the exact analogue of Mertens' result does not hold for double series. (See the examples given on page 1036 of [116] and on page 190 of [22].) The example below shows that the exact analogue of Abel's result does not hold for double series.

Example 9.15 Consider double sequences $(a_{k,\ell})$ and $(b_{k,\ell})$ defined by $a_{0,\ell} := 1$ and $a_{1,\ell} := -1$ for $\ell \in \mathbb{N}_0$, whereas $a_{k,\ell} := 0$ for $k \in \mathbb{N}_0 \setminus \{0, 1\}$ and $\ell \in \mathbb{N}_0$, while $b_{k,0} := 1$ and $b_{k,1} := -1$ for $k \in \mathbb{N}_0$, whereas $b_{k,\ell} := 0$ for $\ell \in \mathbb{N}_0 \setminus \{0, 1\}$ and $k \in \mathbb{N}_0$. Then $\sum_{k,\ell} a_{k,\ell}$ and $\sum_{k,\ell} b_{k,\ell}$ are convergent, and the sum of each is equal to 0. Also, it is easy to see that $a_{0,0} * b_{0,0} = 1$ and $a_{k,\ell} * b_{k,\ell} = 0$ for all $(k, \ell) \in \mathbb{N}_0^2 \setminus \{(0, 0)\}$, so that $\sum_{k,\ell=0}^{\infty} a_{k,\ell} * b_{k,\ell}$ is convergent, and its sum is 1.

We shall now prove analogues of the theorems of Mertens and Abel for double series that are *boundedly convergent*, that is, that are convergent and their partial sums are bounded. In fact, we shall show that Mertens' result admits a converse for such double series in respect of absolute convergence. Our proofs will be based on the following result for a transformation of a double sequence by a 4-fold infinite matrix. It is an analogue of the well-known Kojima–Schur theorem (given, for instance, in Theorem 2.3.7 of [16]) for boundedly convergent double sequences.

Consider a matrix $\mathcal{A} := (\alpha_{m,n,k,\ell})$ with $\alpha_{m,n,k,\ell} \in \mathbb{R}$. We say that $\mathcal{A}$ maps a double sequence $x := (x_{k,\ell})$ to the double sequence $\mathcal{A}x$ defined by

$$[\mathcal{A}x]_{m,n} := \sum_{k,\ell} \alpha_{m,n,k,\ell} x_{k,\ell},$$

provided that the double series on the right side converges for each fixed (m, n).

Lemma 9.16 *A matrix $\mathcal{A} := (\alpha_{m,n,k,\ell})$ maps each bounded convergent double sequence to a bounded convergent double sequence if and only if*

(i) $\sup_{m,n} \sum_{k,\ell} |\alpha_{m,n,k,\ell}| < \infty$,
(ii) *the limit* $\alpha := \lim_{m,n} \sum_{k,\ell} \alpha_{m,n,k,\ell}$ *exists,*
(iii) *the limit* $\alpha_{k,\ell} := \lim_{m,n} \alpha_{m,n,k,\ell}$ *exists for each fixed (k, ℓ), and*
(iv) $\lim_{m,n} \sum_k |\alpha_{m,n,k,\ell} - \alpha_{k,\ell}| = 0$ *for each fixed ℓ,* $\lim_{m,n} \sum_\ell |\alpha_{m,n,k,\ell} - \alpha_{k,\ell}| = 0$ *for each fixed k.*

In this case, the double series $\sum_{k,\ell} \alpha_{k,\ell}$ is absolutely convergent, and for any bounded convergent double sequence $(x_{k,\ell})$, we have the limit formula

$$\lim_{m,n} [\mathcal{A}x]_{m,n} = \left(\alpha - \sum_{k,\ell} \alpha_{k,\ell} \right) \lim_{m,n} x_{m,n} + \sum_{k,\ell} \alpha_{k,\ell} x_{k,\ell}.$$

See [50], especially conditions (c_1), (d_3), and (d_4) in Sect. 3, condition 20. $S.BC \to BC$, limit formula (11.1) in Sect. 6, and a remark in Sect. 7 about the necessity of S. conditions. (See also Theorem 4.1.2 of [127].)

Theorem 9.17 *Let $(a_{k,\ell})$ be a double sequence. Then the Cauchy product double series $\sum_{k,\ell=0}^{\infty} a_{k,\ell} * b_{k,\ell}$ is boundedly convergent for every boundedly convergent double series $\sum_{k,\ell=0}^{\infty} b_{k,\ell}$ if and only if the double series $\sum_{k,\ell=0}^{\infty} a_{k,\ell}$ is absolutely convergent. In this case, we have $\sum_{k,\ell=0}^{\infty} a_{k,\ell} * b_{k,\ell} = (\sum_{k,\ell=0}^{\infty} a_{k,\ell})(\sum_{k,\ell=0}^{\infty} b_{k,\ell})$.*

Proof For $m, n \in \mathbb{N}_0$, let

$$A_{m,n} := \sum_{k=0}^{m} \sum_{\ell=0}^{n} a_{k,\ell}, \qquad B_{m,n} := \sum_{k=0}^{m} \sum_{\ell=0}^{n} b_{k,\ell} \quad \text{and} \quad C_{m,n} := \sum_{k=0}^{m} \sum_{\ell=0}^{n} a_{k,\ell} * b_{k,\ell}.$$

Then

$$C_{m,n} = \sum_{k=0}^{m}\sum_{\ell=0}^{n}\left(\sum_{i=0}^{k}\sum_{j=0}^{\ell}a_{k-i,\ell-j}b_{i,j}\right) = \sum_{k=0}^{m}\sum_{\ell=0}^{n}a_{m-k,n-\ell}\left(\sum_{i=0}^{k}\sum_{j=0}^{\ell}b_{i,j}\right)$$

$$= \sum_{k=0}^{m}\sum_{\ell=0}^{n}a_{m-k,n-\ell}B_{k,\ell} \quad \text{for } (m,n)\in\mathbb{N}_0^2.$$

Now the double series $\sum_{k,\ell=0}^{\infty}a_{k,\ell}*b_{k,\ell}$ (respectively, $\sum_{k,\ell=0}^{\infty}b_{k,\ell}$) is boundedly convergent if and only if the double sequence $(C_{m,n})$ (respectively, $(B_{m,n})$) is boundedly convergent. Consider the matrix $\mathcal{A} := (\alpha_{m,n,k,\ell})$, where

$$\alpha_{m,n,k,\ell} := \begin{cases} a_{m-k,n-\ell} & \text{if } 0\le k\le m \text{ and } 0\le \ell\le n, \\ 0 & \text{otherwise.} \end{cases}$$

Then $[\mathcal{A}(B_{k,\ell})]_{m,n} = C_{m,n}$ for all $(m,n)\in\mathbb{N}_0^2$. It is clear that the matrix $\mathcal{A}$ satisfies condition (i) of Lemma 9.16 if and only if

$$\sup_{m,n}\sum_{k=0}^{m}\sum_{\ell=0}^{n}|a_{m-k,n-\ell}| = \sup_{m,n}\sum_{k=0}^{m}\sum_{\ell=0}^{n}|a_{k,\ell}| < \infty,$$

that is, if and only if the double series $\sum_{k,\ell}a_{k,\ell}$ is absolutely convergent, and in that case, we have

$$\alpha := \lim_{m,n}\sum_{k,\ell}\alpha_{m,n,k,\ell} = \lim_{m,n}\sum_{k=0}^{m}\sum_{\ell=0}^{n}a_{k,\ell} = \sum_{k,\ell}a_{k,\ell},$$

$$\alpha_{k,\ell} := \lim_{m,n}\alpha_{m,n,k,\ell} = \lim_{m,n}a_{m,n} = 0 \quad \text{for each fixed } (k,\ell)\in\mathbb{N}_0^2,$$

$$\lim_{m,n}\sum_{k}|\alpha_{m,n,k,\ell} - \alpha_{k,\ell}| = \lim_{n}\sum_{k=0}^{\infty}|a_{m,n-\ell}| = 0 \quad \text{for each fixed } \ell\in\mathbb{N}_0,$$

$$\lim_{m,n}\sum_{\ell}|\alpha_{m,n,k,\ell} - \alpha_{k,\ell}| = \lim_{m}\sum_{\ell=0}^{\infty}|a_{m-k,n}| = 0 \quad \text{for each fixed } k\in\mathbb{N}_0,$$

that is, conditions (ii), (iii), and (iv) of Lemma 9.16 are automatically satisfied. Hence, the desired result follows from Lemma 9.16 with $\sum_{k,\ell=0}^{\infty}a_{k,\ell}*b_{k,\ell} = (\sum_{k,\ell=0}^{\infty}a_{k,\ell} - 0)\lim_{m,n}B_{m,n} + 0 = (\sum_{k,\ell=0}^{\infty}a_{k,\ell})(\sum_{k,\ell=0}^{\infty}b_{k,\ell})$. $\qquad\square$

Remark 9.18 It is interesting to compare the above analogue of Mertens' theorem with the following result stated by Sheffer in Theorem 3 of [116]. Let $\sum_{k,\ell=0}^{\infty}b_{k,\ell}$ be a convergent double series. Then the Cauchy product double series $\sum_{k,\ell=0}^{\infty}a_{k,\ell}*b_{k,\ell}$ is convergent for every absolutely convergent double series $\sum_{k,\ell=0}^{\infty}a_{k,\ell}$ if and only if the partial sums of $\sum_{k,\ell=0}^{\infty} b_{k,\ell}$ are bounded, and in this case, $\sum_{k,\ell=0}^{\infty} a_{k,\ell} * b_{k,\ell} = (\sum_{k,\ell=0}^{\infty} a_{k,\ell})(\sum_{k,\ell=0}^{\infty} b_{k,\ell})$.

We proceed to prove an analogue of Abel's theorem for boundedly convergent double series. Its proof is based on the following result, which uses the matrix transformation considered in Lemma 9.16.

Lemma 9.19 *Let $(a_{m,n})$ with $(m,n) \in \mathbb{N}_0^2$ be a bounded convergent double sequence and $a := \lim_{m,n} a_{m,n}$. If $\tilde{a}_{m,n} := (\sum_{k=0}^{m} \sum_{\ell=0}^{n} a_{k,\ell})/(m+1)(n+1)$, then $(\tilde{a}_{m,n})$ is a bounded convergent double sequence, and $\lim_{m,n} \tilde{a}_{m,n} = a$. Further, if $(b_{m,n})$ is a bounded convergent double sequence and $b := \lim_{m,n} b_{m,n}$, then*

$$\lim_{m,n} \frac{a_{m,n} * b_{m,n}}{(m+1)(n+1)} = ab.$$

Proof Consider the matrix $\mathcal{L} := (\lambda_{m,n,k,\ell})$, where

$$\lambda_{m,n,k,\ell} := \begin{cases} 1/(m+1)(n+1) & \text{if } 0 \leq k \leq m \text{ and } 0 \leq \ell \leq n, \\ 0 & \text{otherwise.} \end{cases}$$

Then $[\mathcal{L}(a_{k,\ell})]_{m,n} = \tilde{a}_{m,n}$ for all $(m,n) \in \mathbb{N}_0^2$. Also,

$$\sup_{m,n} \sum_{k,\ell} |\lambda_{m,n,k,\ell}| = \lim_{m,n} \sum_{k,\ell} \lambda_{m,n,k,\ell} = \lim_{m,n} \sum_{k=0}^{m} \sum_{\ell=0}^{n} \lambda_{m,n,k,\ell} = 1,$$

$$\lambda_{k,\ell} := \lim_{m,n} \lambda_{m,n,k,\ell} = \lim_{m,n} \frac{1}{(m+1)(n+1)} = 0 \quad \text{for each fixed } (k,\ell) \in \mathbb{N}_0^2,$$

$$\lim_{m,n} \sum_{k} |\lambda_{m,n,k,\ell} - \lambda_{k,\ell}| = \lim_{n} \frac{1}{(n+1)} = 0 \quad \text{for each fixed } \ell \in \mathbb{N}_0, \text{ and}$$

$$\lim_{m,n} \sum_{\ell} |\lambda_{m,n,k,\ell} - \lambda_{k,\ell}| = \lim_{m} \frac{1}{(m+1)} = 0 \quad \text{for each fixed } k \in \mathbb{N}_0,$$

that is, conditions (i)–(iv) of Lemma 9.16 are satisfied, and so by the limit formula, we obtain $\lim_{m,n} \tilde{a}_{m,n} = (1-0)a + 0 = a$.

Next, let $(b_{m,n})$ be a bounded convergent double sequence, and $b := \lim_{m,n} b_{m,n}$. Consider the matrix $\mathcal{A} := (\alpha_{m,n,k,\ell})$, where

$$\alpha_{m,n,k,\ell} := \begin{cases} a_{m-k,n-\ell}/(m+1)(n+1) & \text{if } 0 \leq k \leq m \text{ and } 0 \leq \ell \leq n, \\ 0 & \text{otherwise.} \end{cases}$$

Then $[\mathcal{A}(b_{k,\ell})]_{m,n} = (a_{m,n} * b_{m,n})/(m+1)(n+1)$ for all $(m,n) \in \mathbb{N}_0^2$. Also, since $\beta := \sup\{|a_{m,n}| : (m,n) \in \mathbb{N}_0^2\} < \infty$, we obtain

$$\sup_{m,n} \sum_{k,\ell} |\alpha_{m,n,k,\ell}| = \sup_{m,n} \sum_{k=0}^{m} \sum_{\ell=0}^{n} \alpha_{m,n,k,\ell} \leq \beta,$$

$$\alpha := \lim_{m,n} \sum_{k,\ell} \alpha_{m,n,k,\ell} = \lim_{m,n} \sum_{k=0}^{m} \sum_{\ell=0}^{n} \alpha_{m,n,k,\ell} = \lim_{m,n} \tilde{a}_{m,n} = a,$$

$$\alpha_{k,\ell} := \lim_{m,n} \alpha_{m,n,k,\ell} = \lim_{m,n} \frac{a_{m,n}}{(m+1)(n+1)} = 0 \quad \text{for each fixed } (k, \ell) \in \mathbb{N}_0^2,$$

$$\lim_{m,n} \sum_{k} |\alpha_{m,n,k,\ell} - \alpha_{k,\ell}| \leq \lim_{n} \frac{\beta}{(n+1)} = 0 \quad \text{for each fixed } \ell \in \mathbb{N}_0, \text{ and}$$

$$\lim_{m,n} \sum_{\ell} |\alpha_{m,n,k,\ell} - \alpha_{k,\ell}| \leq \lim_{m} \frac{\beta}{(m+1)} = 0 \quad \text{for each fixed } k \in \mathbb{N}_0,$$

that is, conditions (i)–(iv) of Lemma 9.16 are satisfied, and so by the limit formula we obtain $\lim_{m,n}(a_{m,n} * b_{m,n})/(m+1)(n+1) = (a-0)b + 0 = ab$.   □

A double series $\sum_{k,\ell=0}^{\infty} a_{k,\ell}$ is said to be *Cesàro summable* if

$$\lim_{m,n} \frac{1}{(m+1)(n+1)} \sum_{k=0}^{m} \sum_{\ell=0}^{n} A_{k,\ell}$$

exists, where $A_{k,\ell}$ is the (k, ℓ)th partial sum of the double series. In this case, the above limit is called the *Cesàro sum* of the double series. It follows from Lemma 9.19 that a boundedly convergent double series is Cesàro summable, and its Cesàro sum is equal to its sum.

Theorem 9.20 *Let $\sum_{k,\ell=0}^{\infty} a_{k,\ell}$ and $\sum_{k,\ell=0}^{\infty} b_{k,\ell}$ be boundedly convergent double series. Then the Cauchy product double series $\sum_{k,\ell=0}^{\infty} a_{k,\ell} * b_{k,\ell}$ is Cesàro summable, and its Cesàro sum is equal to AB, where A and B are the sums of the given double series. In particular, if the double series $\sum_{k,\ell=0}^{\infty} a_{k,\ell} * b_{k,\ell}$ is boundedly convergent, then its sum is equal to AB.*

Proof We use the notation introduced in the proof of Theorem 9.17. We have

$$C_{m,n} = \sum_{k=0}^{m} \sum_{\ell=0}^{n} a_{k,\ell} * b_{k,\ell} = a_{m,n} * B_{m,n} = A_{m,n} * b_{m,n} \quad \text{for } (m, n) \in \mathbb{N}_0^2.$$

Replacing $a_{k,\ell}$ and $b_{k,\ell}$ by $A_{m,n}$ and $b_{m,n}$, we obtain

$$\sum_{m=0}^{p} \sum_{n=0}^{q} C_{m,n} = \sum_{m=0}^{p} \sum_{n=0}^{q} A_{m,n} * b_{m,n} = A_{p,q} * B_{p,q} \quad \text{for } (p, q) \in \mathbb{N}_0^2.$$

Hence, by Lemma 9.19 we have

$$\lim_{p,q} \frac{1}{(p+1)(q+1)} \sum_{m=0}^{p} \sum_{n=0}^{q} C_{m,n} = \lim_{p,q} \frac{A_{p,q} * B_{p,q}}{(p+1)(q+1)} = AB.$$

Thus, $\sum_{k,\ell=0}^{\infty} a_{k,\ell} * b_{k,\ell}$ is Cesàro summable, and its Cesàro sum is equal to AB. The last sentence in the statement of the theorem follows easily. $\square$

Remark 9.21   In Theorem 1 of [22], Cesari gives the following analogue of Abel's theorem. Let $\sum_{k,\ell=0}^{\infty} a_{k,\ell}$ and $\sum_{k,\ell=0}^{\infty} b_{k,\ell}$ be convergent double series such that

$$\lim_{k+\ell \to \infty} a_{k,\ell} = 0 = \lim_{k+\ell \to \infty} b_{k,\ell}.$$

Then the double series $\sum_{k,\ell=0}^{\infty} a_{k,\ell} * b_{k,\ell}$ is restrictedly Cesàro summable to AB in the following sense: For any positive real numbers r, s with $r < s$,

$$\lim_{\substack{m,n \to \infty \\ nr < m < ns}} \sigma_{m,n} = AB, \quad \text{where } \sigma_{m,n} := \frac{1}{(m+1)(n+1)} \sum_{k=0}^{m} \sum_{\ell=0}^{n} C_{k,\ell}.$$

We shall now consider a notion of convergence that is stronger than bounded convergence. Let us recall that a double sequence $(a_{k,\ell})$ is said to be *regularly convergent* if it is convergent and, further, for each fixed $k \in \mathbb{N}$, the sequence given by $\ell \mapsto a_{k,\ell}$ is convergent and for each fixed $\ell \in \mathbb{N}$, the sequence given by $k \mapsto a_{k,\ell}$ is convergent. A double series is *regularly convergent* if the double sequence of its (rectangular) partial sums is regularly convergent, that is, the double series is convergent, and further, each corresponding row-series and each corresponding column-series are convergent. It is easy to see that a regularly convergent double sequence is bounded and a regularly convergent double series is boundedly convergent. We have the following analogue of Theorem 9.17 for regularly convergent double series.

Theorem 9.22   *Let $(a_{k,\ell})$ be a double sequence. Then the Cauchy product double series $\sum_{k,\ell=0}^{\infty} a_{k,\ell} * b_{k,\ell}$ is regularly convergent for every regularly convergent double series $\sum_{k,\ell=0}^{\infty} b_{k,\ell}$ if and only if the double series $\sum_{k,\ell=0}^{\infty} a_{k,\ell}$ is absolutely convergent. In this case, we have $\sum_{k,\ell=0}^{\infty} a_{k,\ell} * b_{k,\ell} = (\sum_{k,\ell=0}^{\infty} a_{k,\ell})(\sum_{k,\ell=0}^{\infty} b_{k,\ell})$.*

Proof   An argument along the lines given in the proof of Theorem 9.17 yields the desired result if we note the following. A matrix $\mathcal{A} := (\alpha_{m,n,k,\ell})$ maps each regularly convergent double sequence to a regularly convergent double sequence if and only if conditions (i)–(iii) of Lemma 9.16 hold and also the following condition holds:

(iv)'   the limit $\beta_k := \lim_{m,n} \sum_{\ell} \alpha_{m,n,k,\ell}$ exists for each fixed k, the limit $\gamma_\ell := \lim_{m,n} \sum_{k} \alpha_{m,n,k,\ell}$ exists for each fixed ℓ,

where the convergence indicated in each of conditions (ii), (iii), and (iv)' is regular. In this case, the double series $\sum_{k,\ell} \alpha_{k,\ell}$ is absolutely convergent, the series $\sum_{k} \beta_k$ and $\sum_{\ell} \gamma_\ell$ are absolutely convergent, and for any regularly convergent double sequence $(x_{k,\ell})$, we have the limit formula

$$\lim_{m,n} [\mathcal{A}x]_{m,n} = \left(\alpha + \sum_{k,\ell} \alpha_{k,\ell} - \sum_{k} \beta_k - \sum_{\ell} \gamma_\ell \right) \lim_{m,n} x_{m,n} + \sum_{k,\ell} \alpha_{k,\ell} x_{k,\ell}$$

$$+ \sum_k \left[\left(\beta_k - \sum_\ell \alpha_{k,\ell} \right) \left(\lim_\ell x_{k,\ell} \right) \right]$$

$$+ \sum_\ell \left[\left(\gamma_\ell - \sum_k \alpha_{k,\ell} \right) \left(\lim_k x_{k,\ell} \right) \right].$$

See [50], especially conditions (c_1), (d_1), (d_2), (d_3), (f_1), (f_2), and (f_3) in Sect. 3, condition 132. $S.RC \to RC$, and limit formula (9.1) in Sect. 6. (See also Theorem 4.1.1 of [127].) In our case, with the matrix $\mathcal{A}$ as defined in the proof of Theorem 9.17, we have $\alpha_{k,\ell} = \beta_k = \gamma_\ell = 0$ for all $(k, \ell) \in \mathbb{N}_0^2$ and $\alpha = \sum_{k,\ell} a_{k,\ell}$. □

Remarks 9.23

(i) The "if" part of the above theorem was proved in Theorem 3 of [22]. The "only if" part of the above theorem and of Theorem 9.17 can be strengthened as follows. If the Cauchy product double series $\sum_{k,\ell=0}^\infty a_{k,\ell} * b_{k,\ell}$ is boundedly convergent for every regularly convergent double series $\sum_{k,\ell=0}^\infty b_{k,\ell}$, then the double series $\sum_{k,\ell=0}^\infty a_{k,\ell}$ is absolutely convergent. To see this, one only has to note that condition (i) of Lemma 9.16 is a necessary condition for a matrix $\mathcal{A}$ to map each regularly convergent double sequence to a bounded convergent double sequence. (See condition (c_1) in Sect. 3, condition 134. $S.RC \to BC$ of Sect. 6, and a remark in Sect. 7 about the necessity of S. conditions.)

(ii) In view of (i) above, it is worthwhile to observe that if $\sum_{k,\ell=0}^\infty a_{k,\ell}$ is an absolutely convergent double series not all of whose terms are equal to zero, then there is a boundedly convergent double series $\sum_{k,\ell=0}^\infty b_{k,\ell}$ such that the double series $\sum_{k,\ell=0}^\infty a_{k,\ell} * b_{k,\ell}$ is not regularly convergent. To see this, let $\sum_{k,\ell=0}^\infty a_{k,\ell}$ be absolutely convergent, and let $a_{k_0,\ell_0} \neq 0$ for some $(k_0, \ell_0) \in \mathbb{N}_0^2$. Define $a_k := a_{k,\ell_0}$ for $k \in \mathbb{N}_0$. We note that $\sum_{k=0}^\infty a_k$ is an absolutely convergent series and $a_{k_0} \neq 0$, so that there is a series $\sum_{k=0}^\infty b_k$ having bounded partial sums such that the series $\sum_{k=0}^\infty a_k * b_k$ is divergent. Now define

$$b_{k,\ell} := \begin{cases} b_k & \text{if } k \in \mathbb{N}_0 \text{ and } \ell = 0, \\ -b_k & \text{if } k \in \mathbb{N}_0 \text{ and } \ell = \ell_0 + 1, \\ 0 & \text{if } k \in \mathbb{N}_0 \text{ and } \ell \in \mathbb{N}_0 \setminus \{0, \ell_0 + 1\}. \end{cases}$$

Then

$$a_{k,\ell_0} * b_{k,\ell_0} = \sum_{i=0}^k \sum_{j=0}^{\ell_0} a_{i,j} b_{k-i,\ell_0-j} = \sum_{i=0}^k a_{i,\ell_0} b_{k-i,0} = a_k * b_k.$$

Hence, $\sum_{k=0}^\infty a_{k,\ell_0} * b_{k,\ell_0}$ diverges, and so $\sum_{k,\ell=0}^\infty a_{k,\ell} * b_{k,\ell}$ is not regularly convergent.

Examples 9.24

(i) Let $x, y \in (-1, 0)$ with $x + y = -1$. Define

$$a_{k,\ell} := x^k y^\ell \quad \text{and} \quad b_{k,\ell} := \binom{k+\ell}{k} \quad \text{for } (k, \ell) \in \mathbb{N}_0^2.$$

Then $\sum_{k,\ell=0}^{\infty} a_{k,\ell}$ converges absolutely, and $\sum_{k,\ell=0}^{\infty} b_{k,\ell}$ converges regularly (but not absolutely). The Cauchy product series is $\sum_{k,\ell=0}^{\infty} c_{k,\ell}$, where

$$c_{k,\ell} = \sum_{i=0}^{k} \sum_{j=0}^{\ell} \binom{i+j}{i} x^k y^\ell = \left[\binom{k+\ell+2}{k+1} - 1 \right] x^k y^\ell \quad \text{for } (k, \ell) \in \mathbb{N}_0^2,$$

the last equality being a consequence of Pascal's 3rd identity.

(ii) Let $x, y \in (-1, 1]$, $a, \tilde{a} > 0$, and $b, \tilde{b} \in (-1, 0]$. Define

$$a_{k,\ell} := \binom{a}{k} \binom{\tilde{a}}{\ell} x^k y^\ell \quad \text{and} \quad b_{k,\ell} := \binom{b}{k} \binom{\tilde{b}}{\ell} x^k y^\ell \quad \text{for } (k, \ell) \in \mathbb{N}_0^2.$$

Then $\sum_{k,\ell=0}^{\infty} a_{k,\ell}$ converges absolutely, and $\sum_{k,\ell=0}^{\infty} b_{k,\ell}$ converges regularly (but not absolutely when $x = 1$ or $y = 1$). The Cauchy product series is $\sum_{k,\ell=0}^{\infty} c_{k,\ell}$, where

$$c_{k,\ell} = \sum_{i=0}^{k} \sum_{j=0}^{\ell} \binom{a}{i} \binom{b}{k-i} \binom{\tilde{a}}{j} \binom{\tilde{b}}{\ell-j} x^k y^\ell = \binom{a+b}{k} \binom{\tilde{a}+\tilde{b}}{\ell} x^k y^\ell$$

for $(k, \ell) \in \mathbb{N}_0^2$, the last equality being a consequence of Vandermonde's convolution formula.

(iii) Let $p \in (0, 2]$. Define $a_{k,\ell} := 1/2^{k+\ell}$ and $b_{k,\ell} := (-1)^{k+\ell}/(k + \ell + 1)^p$ for $(k, \ell) \in \mathbb{N}_0^2$. Then $\sum_{k,\ell=0}^{\infty} a_{k,\ell}$ converges absolutely, and $\sum_{k,\ell=0}^{\infty} b_{k,\ell}$ converges regularly (but not absolutely). The Cauchy product series is $\sum_{k,\ell=0}^{\infty} c_{k,\ell}$, where

$$c_{k,\ell} = \sum_{i=0}^{k} \sum_{j=0}^{\ell} \frac{(-1)^{i+j}}{2^{k-i+\ell-j}(i+j+1)^p} = \frac{1}{2^{k+\ell}} \sum_{i=0}^{k} \sum_{j=0}^{\ell} \frac{(-2)^{i+j}}{(i+j+1)^p}$$

for $(k, \ell) \in \mathbb{N}_0^2$.

In each of the above cases, the Cauchy product double series $\sum_{k,\ell=0}^{\infty} c_{k,\ell}$ is regularly convergent by Theorem 9.22.

As we have seen in the beginning of this section, absolute convergence of $\sum_{k,\ell=0}^{\infty} a_{k,\ell}$ and convergence of $\sum_{k,\ell=0}^{\infty} b_{k,\ell}$ do not guarantee convergence of the Cauchy product double series $\sum_{k,\ell=0}^{\infty} a_{k,\ell} * b_{k,\ell}$. In fact, the next result gives precise conditions on the double sequence $(a_{k,\ell})$ for such a convergence.

Theorem 9.25 *Let $(a_{k,\ell})$ be a double sequence. Then the Cauchy product double series $\sum_{k,\ell=0}^{\infty} a_{k,\ell} * b_{k,\ell}$ is convergent for every convergent double series $\sum_{k,\ell=0}^{\infty} b_{k,\ell}$ if and only if the set $\{(k,\ell) \in \mathbb{N}_0^2 : a_{k,\ell} \neq 0\}$ is finite.*

Proof As in the case of Theorem 9.22, an argument along the lines given in the proof of Theorem 9.17 yields the desired result if we note the following. One of the six necessary and sufficient conditions for a matrix $\mathcal{A} := (\alpha_{m,n,k,\ell})$ to map each convergent double sequence to a convergent double sequence is the following: For each fixed $k \in \mathbb{N}_0$, there is $p_0 \in \mathbb{N}_0$ such that $\alpha_{m,n,k,\ell} = 0$ whenever $m, n, \ell \geq p_0$, and for each fixed $\ell \in \mathbb{N}_0$, there is $q_0 \in \mathbb{N}_0$ such that $\alpha_{m,n,k,\ell} = 0$ whenever $m, n, k \geq q_0$. In our case, with the matrix $\mathcal{A}$ as defined in the proof of Theorem 9.17, this condition entails that the set $\{(k,\ell) \in \mathbb{N}_0^2 : a_{k,\ell} \neq 0\}$ is finite, and then the remaining five conditions are automatically satisfied. See [50], especially conditions (a$_1$), (a$_2$), (b$_1$), (b$_2$), (d$_1$), and (d$_3$) in Sect. 3, condition 14. $S.C \to C$ in Sect. 6. (See also Theorem 4.1.3 of [127].) □

Remark 9.26 We conclude this section by making some comments about the companion problem of determining conditions on a double sequence $(a_{k,\ell})$ (with $(k,\ell) \in \mathbb{N}^2$) in order that, for every convergent, boundedly convergent, or regularly convergent double series $\sum_{k,\ell=1}^{\infty} b_{k,\ell}$, the double series $\sum_{k,\ell=1}^{\infty} a_{k,\ell} b_{k,\ell}$ would be convergent, boundedly convergent, or regularly convergent, respectively, as per the assumption on $\sum_{k,\ell=1}^{\infty} b_{k,\ell}$. For this purpose, suppose that a sequence (a_k) is of *bounded variation* if the series $\sum_{k=1}^{\infty} |a_k - a_{k+1}|$ is convergent and that a double sequence $(a_{k,\ell})$ is of *bounded bivariation* if the double series $\sum_{k,\ell=1}^{\infty} |a_{k,\ell} - a_{k+1,\ell} - a_{k,\ell+1} + a_{k+1,\ell+1}|$ is convergent. In the case of regular convergence, necessary and sufficient conditions on $(a_{k,\ell})$ are as follows: $(a_{k,\ell})$ is of bounded bivariation, and $(a_{k,1})$ and $(a_{1,\ell})$ are of bounded variation. In the case of bounded convergence, necessary and sufficient conditions on $(a_{k,\ell})$ are as follows: $(a_{k,\ell})$ is of bounded bivariation, $\lim_k (a_{k,\ell} - a_{k,\ell+1}) = 0$ for each fixed $\ell \in \mathbb{N}$, and $\lim_\ell (a_{k,\ell} - a_{k+1,\ell}) = 0$ for each fixed $k \in \mathbb{N}$. In the case of convergence, necessary and sufficient conditions on $(a_{k,\ell})$ are as follows: $(a_{k,\ell})$ is of bounded bivariation, for each fixed $\ell \in \mathbb{N}$, there is $k_\ell \in \mathbb{N}$ such that $a_{k,\ell} = a_{k,\ell+1}$ for all $k \geq k_\ell$, and for each fixed $k \in \mathbb{N}$, there is $\ell_k \in \mathbb{N}$ such that $a_{k,\ell} = a_{k+1,\ell}$ for all $\ell \geq \ell_k$. See [49] for these results.

References

1. Z.U. Ahmad, M. Mursaleen, An application of Banach limits. Proc. Am. Math. Soc. **103**(1), 244–246 (1988)
2. H. Albayrak, S. Pehlivan, Statistical convergence and statistical continuity on locally solid Riesz spaces. Topol. Appl. **159**, 1887–1893 (2012)
3. A. Alotaibi, C. Çakan, The Riesz convergence and Riesz core of double sequences. J. Inequal. Appl. **2012**, 56 (2012)
4. B. Altay, F. Başar, Some new spaces of double sequences. J. Math. Anal. Appl. **309**, 70–90 (2005)
5. F. Altomare, Korovkin-type theorems and approximation by positive linear operators. Surv. Approx. Theory **5**, 92–164 (2010)
6. G.A. Anastassiou, *Approximation by Multivariate Singular Integrals* (Springer, New York, 2011)
7. G.A. Anastassiou, M. Mursaleen, S.A. Mohiuddine, Some approximation theorems for functions of two variables through almost convergence of double sequences. J. Comput. Anal. Appl. **13**(1), 37–40 (2011)
8. S. Banach, *Théorie des operations liniaries* (Warszawa, 1932)
9. S.B. Baron, Derivation of tests for the convergence of double numerical series. Tartu Riikl. Ül. Toimetised **55**, 9–19 (1958) (in Russian)
10. F. Başar, *Summability Theory and Its Applications* (Bentham Science Publishers, Istanbul, 2011)
11. F. Başar, M. Kirişçi, Almost convergence and generalized difference matrix. Comput. Math. Appl. **61**, 602–611 (2011)
12. M. Başarir, On the strong almost convergence of double sequences. Period. Math. Hung. **30**(3), 177–181 (1995)
13. C. Belen, M. Mursaleen, M. Yildirim, Statistical A-summability of double sequences and a Korovkin type approximation theorem. Bull. Korean Math. Soc. **49**(4), 851–861 (2012)
14. O. Biermann, Über unendliche Doppelreihen und unendliche Doppelproducte. Monatshefte Math. Phys. **8**, 115–137 (1897)
15. G. Bleimann, P.L. Butzer, L. Hahn, A Bernstein type operator approximating continuous functions on semiaxis. Indag. Math. **42**, 255–262 (1980)
16. J. Boos, *Classical and Modern Methods in Summability* (Oxford University Press, Oxford, 2000)
17. B.D. Boyanov, V.M. Veselinov, A note on the approximation of functions in an infinite interval by linear positive operators. Bull. Math. Soc. Sci. Math. Roum. **14**(62), 9–13 (1970)
18. R.C. Buck, *Advanced Calculus*, 3rd edn. (McGraw-Hill, New York, 1978)
19. H. Çakalli, On statistical convergence in topological groups. Pure Appl. Math. Sci. **43**, 27–31 (1996)

M. Mursaleen, S.A. Mohiuddine, *Convergence Methods for Double Sequences and Applications*, DOI 10.1007/978-81-322-1611-7, © Springer India 2014

20. H. Çakalli, E. Savaş, Statistical convergence of double sequence in topological groups. J. Comput. Anal. Appl. **12**(2), 421–426 (2010)

21. Ö. Çakar, A.D. Gadjiev, On uniform approximation by Bleimann, Butzer and Hahn on all positive semiaxis. Trans. Acad. Sci. Azerb. Ser. Phys. Tech. Math. Sci. **19**, 21–26 (1999)

22. L. Cesari, Sulla moltiplicazione delle serie doppie. Ann. Sc. Norm. Super. Pisa, Cl. Sci. **12**, 189–204 (1947)

23. J. Christopher, The asymptotic density of some k-dimensional sets. Am. Math. Mon. **63**, 399–401 (1956)

24. J.S. Connor, The statistical and strong p-Cesàro convergence of sequences. Analysis **8**, 47–63 (1988)

25. R.G. Cooke, *Infinite Matrices and Sequence Spaces* (Macmillan & Co., London, 1950)

26. F. Čunjalo, Almost convergence of double sequences some analogies between measure and category. Math. Maced. **5**, 21–24 (2007)

27. G. Das, Banach and other limits. J. Lond. Math. Soc. (2) **7**, 327–347 (1973)

28. G. Das, B. Kuttner, Space of absolute almost convergence. Indian J. Math. **28**(3), 241–257 (1986)

29. G. Das, S.K. Mishra, A note on a theorem of Maddox on strong almost convergence. Math. Proc. Camb. Philos. Soc. **89**, 393–396 (1981)

30. G. Das, S.K. Sahoo, A generalization of strong and absolute almost convergence. J. Indian Math. Soc. **58**, 65–74 (1992)

31. G. Das, S.K. Sahoo, On some sequence spaces. J. Math. Anal. Appl. **164**, 381–398 (1992)

32. G. Das, B. Kuttner, S. Nanda, Some sequence spaces and absolute almost convergence. Trans. Am. Math. Soc. **283**, 729–739 (1984)

33. G. Das, B. Kuttner, S. Nanda, On absolute almost convergence. J. Math. Anal. Appl. **164**, 381–398 (1992)

34. K. Demirci, F. Dirik, Four-dimensional matrix transformation and rate of A-statistical convergence of periodic functions. Math. Comput. Model. **52**, 1858–1866 (2010)

35. K. Demirci, S. Karakuş, Statistical A-summability of positive linear operators. Math. Comput. Model. **49**, 189–195 (2011)

36. K. Demirci, S. Karakuş, Korovkin-type approximation theorem for double sequences of positive linear operators via statistical A-summability. Results Math. **63**, 1–13 (2013)

37. F. Dirik, K. Demirci, Korovkin-type approximation theorem for functions of two variables in statistical sense. Turk. J. Math. **34**, 73–83 (2010)

38. O. Duman, E. Erkus, Approximation of continuous periodic functions via statistical convergence. Comput. Math. Appl. **52**, 967–974 (2006)

39. J.P. Duran, Infinite matrices and almost convergence. Math. Z. **128**, 75–83 (1972)

40. J.P. Duran, Almost convergence, summability and ergodicity. Can. J. Math. **26**(2), 372–387 (1974)

41. O.H.H. Edely, M. Mursaleen, On statistical A-summability. Math. Comput. Model. **49**, 672–680 (2009)

42. C. Eizen, G. Laush, Infinite matrices and almost convergence. Math. Jpn. **14**, 137–143 (1969)

43. H. Fast, Surla convergence statistique. Colloq. Math. **2**, 241–244 (1951)

44. A.P. Freedman, J.J. Sember, Densities and summability. Pac. J. Math. **95**, 293–305 (1981)

45. J.A. Fridy, On statistical convergence. Analysis **5**, 301–313 (1985)

46. J.A. Fridy, Statistical limit points. Proc. Am. Math. Soc. **118**, 1187–1192 (1993)

47. J.A. Fridy, C. Orhan, Statistical limit superior and limit inferior. Proc. Am. Math. Soc. **125**, 3625–3631 (1997)

48. M. Gurdek, L. Rempulska, M. Skorupka, The Baskakov operators for functions of two variables. Collect. Math. **50**, 289–302 (1999)

49. H.J. Hamilton, On transformations of double series. Bull. Am. Math. Soc. **42**, 275–283 (1936)

50. H.J. Hamilton, Transformations of multiple sequences. Duke Math. J. **2**, 29–60 (1936)

51. S. Karakus, K. Demirci, Statistical convergence of double sequences on probabilistic normed spaces. Int. J. Math. Math. Sci. **2007**, ID 14737 (2007), 11 pp.

52. S. Karakus, K. Demirci, O. Duman, Statistical convergence on intuitionistic fuzzy normed spaces. Chaos Solitons Fractals **35**, 763–769 (2008)
53. K. Kayaduman, C. Çakan, The Cesàro core of double sequences. Abstr. Appl. Anal. **2011**, ID 950364 (2011), 9 pp.
54. M. Khan, C. Orhan, Matrix characterization of A-statistical convergence. J. Math. Anal. Appl. **335**(1), 406–417 (2007)
55. J.P. King, Almost summable sequences. Proc. Am. Math. Soc. **17**, 1219–1225 (1966)
56. K. Knopp, Zur Theorie der Limitierungsverfahren (Erste Mitteilung). Math. Z. **31**, 115–127 (1930)
57. E. Kolk, The statistica convergence in Banach spaces. Tartu Ülik Toimetised **928**, 41–52 (1991)
58. E. Kolk, Matrix summability of statistically convergent sequences. Analysis **13**, 77–83 (1993)
59. P.P. Korovkin, Convergence of linear positive operators in the spaces of continuous functions. Dokl. Akad. Nauk SSSR (N.S.) **90**, 961–964 (1953) (Russian)
60. P.P. Korovkin, *Linear Operators and Approximation Theory* (Hindustan Publ. Co., Delhi, 1960)
61. V. Kumar, M. Mursaleen, On (λ, μ)-statistical convergence of double sequences on intuitionistic fuzzy normed spaces. Filomat **25**(2), 109–120 (2011)
62. B.V. Limaye, M. Zeltser, On the Pringsheim convergence of double series. Proc. Est. Acad. Sci., Phys. Math. **58**(2), 108–121 (2009)
63. G.G. Lorentz, A contribution to theory of divergent sequences. Acta Math. **80**, 167–190 (1948)
64. I.J. Maddox, A new type of convergence. Math. Proc. Camb. Philos. Soc. **83**, 61–64 (1978)
65. I.J. Maddox, On strong almost convergence. Math. Proc. Camb. Philos. Soc. **85**, 345–350 (1979)
66. I.J. Maddox, Some analogues of Knopp's core theorem. Int. J. Math. Math. Sci. **2**, 605–614 (1979)
67. I.J. Maddox, Statistical convergence in a locally convex space. Math. Proc. Camb. Philos. Soc. **104**, 141–145 (1988)
68. G.D. Maio, L.D.R. Kočinac, Statistical convergence in topology. Topol. Appl. **156**, 28–45 (2008)
69. F.M. Mears, Absolute regularity and the Nörlund mean. Ann. Math. **38**, 594–601 (1937)
70. W. Meyer-König, K. Zeller, Bersteinsche Potenzreihen. Stud. Math. **19**, 89–94 (1960)
71. H.I. Miller, A measure theoretical subsequence characterization of statistical convergence. Trans. Am. Math. Soc. **347**(5), 1811–1819 (1995)
72. R.N. Mohapatra, Quantitative results on almost convergence of a sequence of positive linear operators. J. Approx. Theory **20**, 239–250 (1977)
73. S.A. Mohiuddine, An application of almost convergence in approximation theorems. Appl. Math. Lett. **24**, 1856–1860 (2011)
74. S.A. Mohiuddine, M. Aiyub, Lacunary statistical convergence in random 2-normed spaces. Appl. Math. Inf. Sci. **6**(3), 581–585 (2012)
75. S.A. Mohiuddine, M.A. Alghamdi, Statistical summability through a lacunary sequence in locally solid Riesz spaces. J. Inequal. Appl. **2012**, 225 (2012)
76. S.A. Mohiuddine, A. Alotaibi, Statistical convergence and approximation theorems for functions of two variables. J. Comput. Anal. Appl. **15**(2), 218–223 (2013)
77. S.A. Mohiuddine, Q.M. Danish Lohani, On generalized statistical convergence in intuitionistic fuzzy normed space. Chaos Solitons Fractals **42**, 1731–1737 (2009)
78. S.A. Mohiuddine, E. Savaş, Lacunary statistically convergent double sequences in probabilistic normed spaces. Ann. Univ. Ferrara **58**, 331–339 (2012)
79. S.A. Mohiuddine, A. Alotaibi, M. Mursaleen, Statistical summability $(C, 1)$ and a Korovkin type approximation theorem. J. Inequal. Appl. **2012**, 172 (2012)
80. S.A. Mohiuddine, A. Alotaibi, M. Mursaleen, Statistical convergence of double sequences in locally solid Riesz spaces. Abstr. Appl. Anal. **2012**, ID 719729 (2012), 9 pp.

81. F. Móricz, Tauberian theorems for Cesàro summable double sequences. Stud. Math. **110**, 83–96 (1994)
82. F. Móricz, Statistical convergence of multiple sequences. Arch. Math. **81**, 82–89 (2003)
83. F. Móricz, B.E. Rhoades, Almost convergence of double sequences and strong regularity of summability matrices. Math. Proc. Camb. Philos. Soc. **104**, 283–294 (1988)
84. M. Mursaleen, On some new invariant matrix methods of summability. Q. J. Math. (Oxford) **34**, 77–86 (1983)
85. M. Mursaleen, Matrix transformations between some new sequence spaces. Houst. J. Math. **9**, 505–509 (1983)
86. M. Mursaleen, Infinite matrices and absolute almost convergence. Int. J. Math. Math. Sci. **6**, 503–510 (1983)
87. M. Mursaleen, Absolute almost convergent sequences. Houst. J. Math. **10**(3), 427–431 (1984)
88. M. Mursaleen, Almost strongly regular matrices and a core theorem for double sequences. J. Math. Anal. Appl. **293**, 523–531 (2004)
89. M. Mursaleen, Some matrix transformations on sequence spaces of invariant means. Hacet. J. Math. Stat. **38**(3), 259–264 (2009)
90. M. Mursaleen, On statistical convergence in random 2-normed spaces. Acta Sci. Math. (Szeged) **76**, 101–109 (2010)
91. M. Mursaleen, On $\mathcal{A}$-invariant mean and $\mathcal{A}$-almost convergence. Anal. Math. **37**(3), 173–180 (2011)
92. M. Mursaleen, A. Alotaibi, On I-convergence in random 2-normed spaces. Math. Slovaca **61**(6), 933–940 (2011)
93. M. Mursaleen, O.H.H. Edely, Statistical convergence of double sequences. J. Math. Anal. Appl. **288**, 223–231 (2003)
94. M. Mursaleen, O.H.H. Edely, Almost convergence and a core theorem for double sequences. J. Math. Anal. Appl. **293**, 532–540 (2004)
95. M. Mursaleen, O.H.H. Edely, Generalized statistical convergence. Inf. Sci. **162**, 287–294 (2004)
96. M. Mursaleen, S.A. Mohiuddine, Almost bounded variation of double sequences and some four dimensional summability matrices. Publ. Math. (Debr.) **75**, 495–508 (2009)
97. M. Mursaleen, S.A. Mohiuddine, Statistical convergence of double sequences in intuitionistic fuzzy normed spaces. Chaos Solitons Fractals **41**, 2414–2421 (2009)
98. M. Mursaleen, S.A. Mohiuddine, On lacunary statistical convergence with respect to the intuitionistic fuzzy normed space. J. Comput. Appl. Math. **233**, 142–149 (2009)
99. M. Mursaleen, S.A. Mohiuddine, On ideal convergence of double sequences in probabilistic normed spaces. Math. Rep. **12**(64)(4), 359–371 (2010)
100. M. Mursaleen, S.A. Mohiuddine, Banach limit and some new spaces of double sequences II. J. Indian Math. Soc. **78**, 117–130 (2011)
101. M. Mursaleen, S.A. Mohiuddine, Banach limit and some new spaces of double sequences. Turk. J. Math. **36**, 121–130 (2012)
102. M. Mursaleen, S.A. Mohiuddine, On ideal convergence in probabilistic normed spaces. Math. Slovaca **62**(1), 49–62 (2012)
103. M. Mursaleen, E. Savaş, Almost regular matrices for double sequences. Studia Sci. Math. Hung. **40**, 205–212 (2003)
104. M. Mursaleen, C. Çakan, S.A. Mohiuddine, E. Savaş, Generalized statistical convergence and statistical core of double sequences. Acta Math. Sin. Engl. Ser. **26**, 2131–2144 (2010)
105. M. Mursaleen, S.A. Mohiuddine, O.H.H. Edely, On the ideal convergence of double sequences in intuitionistic fuzzy normed spaces. Comput. Math. Appl. **59**, 603–611 (2010)
106. T. Neubrum, J. Smital, T. Šalát, On the structure of the space $M(0, 1)$. Rev. Roum. Math. Pures Appl. **13**, 337–386 (1968)
107. R.H. Patterson, Double sequence core theorems. Int. J. Math. Math. Sci. **22**, 785–793 (1999)
108. A. Pringsheim, Zur theorie der zweifach unendlichen Zahlenfolgen. Math. Z. **53**, 289–321 (1900)

109. R.A. Raimi, Invariant means and invariant matrix methods of summability. Duke Math. J. **30**, 81–94 (1963)

110. G.T. Roberts, Topologies in vector lattices. Math. Proc. Camb. Philos. Soc. **48**, 533–546 (1952)

111. G.M. Robinson, Divergent double sequences and series. Trans. Am. Math. Soc. **28**, 50–73 (1926)

112. T. Šalát, On statistically convergent sequences of real numbers. Math. Slovaca **30**, 139–150 (1980)

113. E. Savaş, S.A. Mohiuddine, $\bar{\lambda}$-statistically convergent double sequences in probabilistic normed spaces. Math. Slovaca **62**(1), 99–108 (2012)

114. P. Schaefer, Infinite matrices and invariant means. Proc. Am. Math. Soc. **36**, 104–110 (1972)

115. I.J. Schoenberg, The integrability of certain functions and related summability methods. Am. Math. Mon. **66**, 361–375 (1959)

116. I.M. Sheffer, Note on multiply-infinite series. Bull. Am. Math. Soc. **52**, 1036–1041 (1946)

117. S. Simons, The sequence spaces $l(p_\nu)$ and $m(p_\nu)$. Proc. Lond. Math. Soc. (3) **15**, 422–436 (1965)

118. S. Simons, Banach limits, infinite matrices and sublinear functionals. J. Math. Anal. Appl. **26**, 640–655 (1969)

119. D.D. Stancu, A method for obtaining polynomials of Bernstein type of two variables. Am. Math. Mon. **70**(3), 260–264 (1963)

120. H. Steinhaus, Sur la convergence ordinaire et la convergence asymptotique. Colloq. Math. **2**, 73–74 (1951)

121. M. Stieglitz, Eine Verallgemeinerung des Begriffs der Fastkonvergenz. Math. Jpn. **18**, 53–70 (1973)

122. F. Taşdelen, A. Erençin, The generalization of bivariate MKZ operators by multiple generating functions. J. Math. Anal. Appl. **331**, 727–735 (2007)

123. B.C. Tripathy, Statistically convergent double sequences. Tamkang J. Math. **34**(3), 231–237 (2003)

124. V.I. Volkov, On the convergence of sequences of linear positive operators in the space of two variables. Dokl. Akad. Nauk SSSR **115**, 17–19 (1957)

125. N.N. Vorob'ev, *Theory of Series*, 5th edn. (Nauka, Moscow, 1986) (in Russian)

126. A.C. Zaanen, *Introduction to Operator Theory in Riesz Spaces* (Springer, Heidelberg, 1997)

127. M. Zeltser, Investigation of double sequence spaces by soft and hard analytical methods, Ph.D. thesis, Tartu Univ. Press, Tartu, 2001

128. M. Zeltser, M. Mursaleen, S.A. Mohiuddine, On almost conservative matrix methods for double sequence spaces. Publ. Math. (Debr.) **75**, 387–399 (2009)

Printed in the United States
By Bookmasters

Printed in the United States
By Bookmasters